CHEMISTRY AND PHYSICS
OF CARBON

Volume 20

CHEMISTRY AND PHYSICS OF CARBON

A SERIES OF ADVANCES

Edited by

Peter A. Thrower

DEPARTMENT OF MATERIALS SCIENCE AND ENGINEERING
THE PENNSYLVANIA STATE UNIVERSITY
UNIVERSITY PARK, PENNSYLVANIA

Volume 20

CRC Press
Taylor & Francis Group
Boca Raton London New York

CRC Press is an imprint of the
Taylor & Francis Group, an **informa** business

First Published 1987 by MARCEL DEKKER, INC.

Published 2021 by CRC Press
Taylor & Francis Group
6000 Broken Sound Parkway NW, Suite 300
Boca Raton, FL 33487-2742

© 1987 by Taylor & Francis Group, LLC
CRC Press is an imprint of Taylor & Francis Group, an Informa business

No claim to original U.S. Government works

ISBN 13: 978-0-8247-7740-1 (hbk)
ISBN 13: 978-0-367-45152-3 (pbk)
ISBN 13: 978-1-00-320903-4 (ebk)

DOI: 10.1201/9781003209034

**Visit the Taylor & Francis Web site at
http://www.taylorandfrancis.com**

**and the CRC Press Web site at
http://www.crcpress.com**

The Library of Congress Cataloged the
First Issue of This Title as Follows:

Chemistry and physics of carbon, v. 1-
    London, E. Arnold; New York, M. Dekker, 1965-

       v. illus. 24 cm

    Editor: v. 1- P. L. Walker

    1. Carbon. I. Walker, Philip L., ed.

QD181.C1C44   546.681

Library of Congress   1   66-58302
ISBN 0-8247-7740-9

Preface

By the time this volume of *Chemistry and Physics of Carbon* is published, over two years will have elapsed since the publication of
Volume 19. Such are the hazards of compiling volumes of this nature.
The science of carbon materials develops and expands with each passing
month and year and few people are willing to sacrifice the time and
effort to place recent developments in the perspective required for
a volume such as this. The first chapter in this volume, "Structural
Studies of PAN-Based Carbon Fibers," was submitted to the editor by
David Johnson over two years ago. I am grateful to the author for
having taken the time to revise the manuscript to bring it up-to-date
for this volume. A chapter on rayon-based fibers by Roger Bacon
appeared in Volume 9 of this series, and I trust that a chapter on
pitch-based fibers will eventually be available. Please contact me
if you wish to volunteer!

    The second chapter is also the product of a long gestation
period. Chapters on the electronic properties of carbons have
appeared in several earlier volumes of this series and the subject
is of perennial interest. Dr. Marie-France Charlier and Dr. Alphonse
Charlier have attempted in this contribution to provide a fundamental
understanding of the basis of the theoretical treatment of electronic
properties in graphite. I am particularly grateful to Dr. John
Woollam of the University of Nebraska for his help in arriving at
the final version of the manuscript.

Graphite intercalation compounds have been the subject of intense interest in recent years. Many species are able to enter between the layer planes and the variety of compounds produced is quite remarkable. The authors of the third chapter have focused their attention on the alkali metals, sodium and potassium, whose interactions with graphite have practical importance in the blast furnace and in aluminum production. The chapter concludes with an assessment of the available literature on this topic and an examination of the mechanism of potassium intercalation.

This volume again illustrates the wide range of topics of interest to researchers on carbon materials and I trust will stimulate further understanding of some of the phenomena involved.

Peter A. Thrower

Contributors to Volume 20

Hanns-Peter Boehm Institut fur Anorganische Chemie, Universitat München, München, Federal Republic of Germany

Alphonse Charlier Faculté des Sciences, Université de Metz, Metz, France

Marie-France Charlier Faculté des Sciences, Université de Metz, Metz, France

Ian A. S. Edwards Northern Carbon Research Laboratories, University of Newcastle upon Tyne, Newcastle upon Tyne, United Kingdom

David J. Johnson Department of Textile Industries, University of Leeds, Leeds, United Kingdom

Harry Marsh Northern Carbon Research Laboratories, University of Newcastle upon Tyne, Newcastle upon Tyne, United Kingdom

Neil Murdie[*] Northern Carbon Research Laboratories, University of Newcastle upon Tyne, Newcastle upon Tyne, United Kingdom

---

[*]*Present affiliation*: Materials Technology Center, Southern Illinois University, Carbondale, Illinois

Contents of Volume 20

*Harry Marsh, Neil Murdie, Ian A. S. Edwards
and Hanns-Peter Boehm*

Contents of Other Volumes

CHEMISTRY AND PHYSICS OF CARBON

Volume 20

1

Structural Studies of PAN-Based Carbon Fibers

DAVID J. JOHNSON

Dept. of Textile Industries, University of Leeds
Leeds, United Kingdom

I. INTRODUCTION

In the years since Reynolds reviewed the structure and physical
properties of carbon fibers [1], profound changes have taken place,
in particular the successful development of carbon fibers from mesophase pitch [2,3] and the decline of carbon fibers from rayon pre-
cursors [4]. There have been very significant improvements in the
performance of both type I and type II PAN-based carbon fibers,
together with the introduction of both ultra-high-modulus type I and
high-strength type A fibers, with the result that there is a wide
variety of commercially available carbon fibers suitable for many
different end uses. During this period, and despite predictions to
the contrary, PAN-based fibers have maintained their predominance as
a reinforcing material; their continued success can be attributed to
inherent structural characteristics that ensure the most beneficial
physical properties for that purpose. The basic physical properties
of several different carbon fibers are given in Table 1; the values
of Young's modulus and tensile strength are well in excess of those
considered possible only a few years ago and reflect the advance of
an important high-technology material no longer searching for suitable end uses.

    The engineer working with high-performance materials has the
requirement of high modulus, high strain-to-failure, high tensile
strength, and, as far as possible, easy handling. High modulus has
been relatively straightforward to achieve, but usually at the
expense of low strain-to-failure and brittleness. To accomplish
both high modulus and high strain-to-failure, which, because of the
straight-line stress-strain relationship means high tensile strength,
has been more difficult. Although it has long been recognized that
the optimum development of tensile strength may be hindered by the
presence of gross defects, we have only recently seen significant
improvement in the realized tensile strengths of carbon fibers as
these defects have been reduced or eliminated. Unfortunately, high
tensile strength and ultra-high modulus are all too often accompanied
by brittleness and low flexibility. However, in this respect, most

TABLE 1 Tensile Properties of Carbon Fibers (Manufacturers' Data)

Manufacturer	Fiber	Young's modulus E (GPa)	Tensile strength σ (GPa)	Strain to failure ε (%)
PAN-based type I				
Courtaulds	HM-S Grafil	340	2.10	0.6
Hercules	HM-S Magnamite	345	2.21	0.6
Celanese	Celion GY-70	517	1.86	0.4
Toray	M-50	500	2.50	0.5
PAN-based type II				
Toho Beslon	Sta-grade Besfight	240	3.73	1.6
Union Carbide	Thornel 300	230	3.10	1.3
Celanese	Celion 1000	234	3.24	1.4
Hercules	IM-6	276	4.48	1.6
Toray	T-800	300	5.70	1.9
PAN-based type A				
Courtaulds	XA-S Grafil	230	2.90	1.3
Celanese	Celion ST	235	4.34	1.8
Hercules	AS-6	241	4.14	1.7
Mesophase pitch-based				
Union Carbide	Thornel P-25	140	1.40	1.0
	P-55	380	2.10	0.5
	P-75	500	2.00	0.4
	P-100	690	2.20	0.3
	P-120	820	2.20	0.2

PAN-based carbon fibers are rather more resistant to the compressive forces applied in flexure than liquid crystal-based materials, such as the carbon fibers from thermotropic mesophase pitches or the high-performance organic fibers from lyotropic polymers, for example, PPT (poly *p*-phenylene terephthalamide), which is marketed as Kevlar.[*]

Over the years it has been found that the most effective techniques to determine structure in carbon fibers are x-ray diffraction, both wide- and small-angle, and electron microscopy, both scanning and transmission. Much of the earlier work using these techniques

---

[*]Kevlar is the registered trademark of E. I. duPont de Nemours & Company, Inc.

was aimed at straightforward characterization of the materials
studied, but although less emphasis was placed on the relationships
between structure and properties, it was the inherent need to exploit
these relationships that initiated those first investigations. We
have now arrived at the situation in which type I PAN-based carbon
fibers have been well characterized and we have considerable knowledge of structural features and failure mechanisms that limit tensile
strength. How far this understanding applies to type II and type A
fibers is open to question; nevertheless, improved all-around understanding has prompted a considerable increase in the strain-to-failure
and tensile strength of available fibers. At the same time, it is
reasonable to suggest that much less attention has been paid to deformations that involve compression or combined compression and tension
in carbon fibers of all origins and that the mechanisms of failure in
these deformation modes are less well understood.

Here, we briefly summarize the essential structures of PAN-based
carbon fibers as discovered by the application of x-ray diffraction
and electron microscopy and show how the physical properties are related to the inherent structure. This treatment is selective rather
than comprehensive but provides a foundation for a discussion of more
recent work, which will be dealt with in greater detail and will draw
on studies of both mesophase pitch-based carbon fibers and PPT fibers
for comparison.

II.   X-RAY DIFFRACTION

A.   Wide-angle X-ray Diffraction

Wide-angle x-ray diffraction is concerned with the scatter of x-rays
by atoms in a material, reflections occurring where the scattered
x-rays are in phase. Bragg's law, $n\lambda = 2d \sin \theta$, where λ is the
wavelength, d the interplanar spacing, and θ half the angle of scattering, is used to relate the position of an x-ray diffraction peak
to the interplanar spacing. Wide-angle (or high-angle) x-ray diffraction is the simplest and quickest technique for the structural
characterization of any material and, in terms of carbon, has been

reviewed in detail by Ruland [5]. Accounts more specific to the
analysis of carbon fiber patterns have also been published [6,7].

Typical wet-spun PAN fibers have two main reflections on the
equator, no reflections on the meridian, and no layer-line reflections. The orientation of the molecular chains is low, and there
is no true unit cell, only pseudohexagonal packing at a spacing of
about 0.51 nm. After prestretching, there is a great improvement in
orientation (Fig. 1), and in the original diffraction pattern, it is
possible to see first layer-line scattering. True three-dimensional
order of the molecules in PAN fibers cannot be achieved, even by
stretching.

After carbonization above 1000°C, carbon fibers exhibit a typical
x-ray diffraction pattern with a broad 002 reflection on the equator
and a diffuse 100 ring, strongest on the meridian (Fig. 2). The small-
angle scatter can also be seen but, there are no three-dimensional
reflections as found in graphite. After heat treatment at 2500°C,
type I carbon fibers show much sharper 002 and 100 reflections, and
the 004 reflection is evident on the equator (Fig. 3); again, there
are no reflections characteristic of three dimensional order, such
as 101 or 112. Essentially, PAN-based carbon fibers are nongraphitizing with a turbostratic organization of the layer planes. In contrast, mesophase pitch-based carbon fibers are graphitizing, and

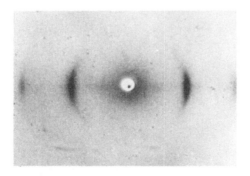

FIG. 1 Wide-angle x-ray diffraction pattern of PAN fibers
stretched ×12.

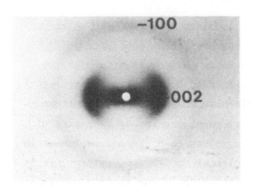

FIG. 2 Wide-angle x-ray diffraction pattern of type II PAN-based
carbon fiber.

high-temperature treated fibers show evidence of a strong 101 reflection outside the 100 ring (Fig. 4), indicating the more perfect
arrangement of the layer planes.

Although much information can be gained by a qualitative study
of the x-ray diffraction pattern, serious characterization necessi-
tates more rigorous quantitative analysis. By means of an x-ray
diffractometer, usually employed in step-scan mode, it is possible
to obtain intensity data on the diffraction peaks in both equatorial
and meridional directions. The measure of peak width B_{002} from the

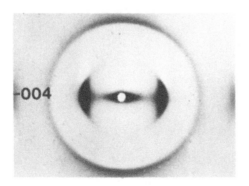

FIG. 3 Wide-angle x-ray diffraction pattern of type I PAN-based
carbon fiber.

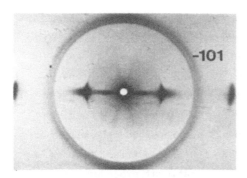

FIG. 4 Wide-angle x-ray diffraction pattern from heat-treated
mesophase pitch-based carbon fiber.

002 reflection is used to measure the stacking size L_c by means of
the relationship $L_c = K\lambda/B_{002} \cos \theta$, where θ is the Bragg angle and
λ the x-ray wavelength. The Scherrer parameter K can be considered
the factor by which the apparent size must be multiplied to determine
the true size, and for 00*l* reflections is generally given the value
of 1. The 100 reflection is spread into a ring because of the turbostratic packing; the meridional width is used to obtain the layer-
plane length parallel to the fiber axis $L_{a\parallel}$; similarly, the equatorial
width is used to obtain the layer-plane dimension perpendicular to the
fiber axis $L_{a\perp}$. Both these measures should be considered essentially
"correlation lengths" relating only to short perfect distances in the
fiber as "seen" by the x-rays. In fact, transmission electron microscopy shows that the layer-plane lengths and widths are very much
greater than the L_a values measured by x-ray diffraction. K values
for hk0 reflections (traditionally 1.84 for $L_{a\perp}$) have been evaluated
by Ruland and Tompa [8] as a function of the preferred orientation.

An x-ray diffraction trace around a 00*l* peak in an azimuthal
direction can be used to obtain quantitative information concerning
the preferred orientation of the layer planes. Two useful parameters
are Z, the azimuthal width at half-peak height, and q, the orientation
factor introduced by Ruland and Tompa [9]. This factor is somewhat
complex mathematically, but in essence, a fiber with no orientation

will have a q factor of 0, and a fiber with layer planes oriented
perfectly along its axis will have a q factor of -1.

When the diffraction peaks overlap, it is necessary to separate
them by computational methods. A full account of all the steps required to resolve the peaks and evaluate crystallite size is given
in Ref. 10, and a short account specific to carbon fibers is given
in Ref. 11. In brief, overlapping peaks are resolved into mixed
Gaussian-Cauchy profiles, which may be either symmetrical or asymmetric about the peak position. The computational procedure is a
minimization of the sum of squares between the observed and the calculated data in terms of parameters defining peak-profile shape, peak
height, peak width, and peak position. Parameters can be constrained
within set limits when necessary. Each $00l$ peak is a convolution of
size-broadening and distortion-broadening components. Numerous mathematical techniques have been proposed for the separation of size and
disorder using both transform and nontransform methods (see Ref. 11)
in which results from real and simulated profiles are given after
evaluation by several different routes. Typical values of x-ray
diffraction parameters for various types of PAN-based carbon fibers
are given in Table 2. It should be noted that the lattice disorder
σ_l, which is a root mean square value for the lattice distortion $\Delta d/d$

TABLE 2 Structural Parameters of PAN-based Carbon Fibers Obtained
by X-ray Diffraction[a]

Fiber	c/2 (nm)	L_c (nm)	$L_{a\parallel}$ (nm)	$L_{a\perp}$ (nm)	σ_l	Z (°)	-q	l_p (nm)
Type I	0.35	5.3	9.8	8.0	0.9	20	0.78	2.5
Type II	0.36	1.7	3.9	2.7	—	41	0.55	1.4
Type A	0.37	1.2	3.2	1.9	—	44	0.51	0.9

[a]L_c crystallite size in c direction (stacking size). $L_{a\parallel}$ crystallite
size in a direction parallel to fiber axis. $L_{a\perp}$ crystallite size in
a direction perpendicular to fiber axis. σ_l lattice order (cannot be
measured for type II and type A fibers). Z preferred orientation.
q preferred orientation factor. l_p pore size perpendicular to fiber
axis.

cannot be measured for type II or type A fibers because there is only one 00l reflection.

B. Small-angle X-ray Diffraction

When structural units in a material scatter x-rays, they do so at small angles of scatter and special instrumental techniques are necessary to obtain information about this small- (or low-) angle diffraction. If discrete reflections are observed in the small-angle pattern, then the repeat of the structure l can be evaluated by the application of Bragg's law, although this must be considered as a first approximation that can be fully solved by more rigorous mathematics. It is best to consider the long spacing l as a measure of a periodic density difference.

We may note that many thermoplastic fibers, such as poly(ethylene terephthalate) (PET), have a broad meridional reflection (see Fig. 5a), which can be interpreted as a long spacing of about 10 nm arising from the chain-folded nature of the molecules [12]. The natural protein

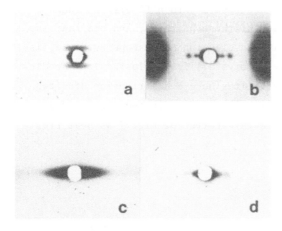

FIG. 5 Small-angle x-ray diffraction patterns of fibers. (a) Poly (ethylene terephthalate) (PET), an example of a chain-folded structure. (b) Lincoln wool, a fiber with a discrete fibrillar structure. (c) Fortisan, a regenerated cellulose fiber with a range of fibrillar size. (d) PAN; this example shows evidence of both discrete and non-discrete fibrils.

fiber, wool, has discrete equatorial reflections at 8.0, 4.5, and
2.7 nm (Fig. 5b), due to a microfibrillar arrangement of protofibrils
and their packed structure of fundamental coiled-coil keratin molecules [13]. Other fibers, for example, ramie and Fortisan, the highly
crystalline fibers of native and regenerated cellulose, do not have
discrete reflections; they exhibit a streak along the equator (Fig.
5c), which indicates that the structure is one of fibrils or crystallites with a distribution of widths [14]. A typical PAN fiber has a
very weak equatorial streak (Fig. 5d), which, in this case, indicates
the presence of both discrete fibrils together with some larger non-
discrete fibrillar structures.

Carbon fibers have a very strong equatorial lobe-shaped pattern,
the development of which is illustrated in Fig. 6, in which small-
angle patterns recorded from fibers heat treated to different temperatures are shown. The faint discrete spot and equatorial streak from
the precursor PAN changes to a more intense streak with increasing
temperature and is gradually replaced by the lobe-shaped carbon fiber
pattern. Small-angle patterns showing the full development of the
lobe-shaped scatter in type II and type I carbon fibers are shown in
Fig. 7. The best interpretation of these patterns, and one that is
in good agreement with transmission electron microscope evidence, is
that there is a fine structure of crystallites enclosing long needle-
shaped voids and that the crystallites have a distribution of sizes.
The effect of improved orientation can be seen in the decreased spread
of the lobe-shaped small-angle x-ray scatter of Fig. 7 as heat treatment temperature is increased.

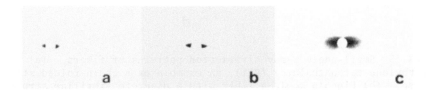

FIG. 6 Small-angle x-ray diffraction patterns from PAN fibers heat
treated to different temperatures: (a) 300°C, (b) 500°C, and (c)
700°C.

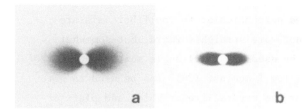

FIG. 7 Small-angle x-ray diffraction patterns from PAN-based carbon fibers: (a) type II; (b) type I.

If it is necessary to obtain a measure of the void or pore size for a powder specimen, a "mean chord intercept length" in the pores l_p can be measured using the Debye method [15,16]. A mean chord intercept length in the crystallites l_c can also be evaluated. Plots of $I^{-1/2}$ against θ^2 are made and the slope-intercept ratio found to give the so-called correlation length a. If the ratio of the density of the fiber to that of perfect graphite c is known, then the correlation length can be used to find l_p, l_c, and the specific surface S_v. The relationships used are:

$$l_p = 2a$$
$$l_c = a(1 - c$$
$$S_v = \frac{4c}{l_c}$$

Although reasonable approximations can be made for other fibers [16], these relationships are only completely valid for type I fibers.

The distinction between a sharp boundary from crystallite to pore and a boundary that might contain disordered layer planes can be tested by Porod's law. For a powder, this is that a plot of $I\theta^4$ against 2θ reaches a constant limit for a sharp boundary and oscillates about that limit for a boundary containing disordered layers. For a fiber specimen, the law requires an $I\theta^3$ against 2θ relationship. Perret and Ruland [17,18] showed that it is better to plot $I\theta^3$ against $(2\theta)^2$ for fibers. In this case, the plots tend to a straight line and do not oscillate; the part of the curve due to disordered layers can then be removed. By suitable corrections and normalization, values of l_p can be obtained.

Values of l_p measured perpendicular to the fiber axis are
included in Table 2. Other more straightforward, but somewhat
approximate, methods can be used for small-angle scattering, for
example, Guinier plots of log I against $(2\theta)^2$ can be used to give
a radius of gyration R for the crystallite or void, and plots of
log I against $\log \theta^2$ can be used to give parameters for the distribution of void size [19].

III. ELECTRON MICROSCOPY

A. Scanning Electron Microscopy

The scanning electron microscope (SEM) is the workhorse microscope
in materials science; results are obtained rapidly and usually have
sufficient information content for deductions of structural signifi-
cance. Also, additional instrumentation is often attached to the
SEM, for example for elemental analysis using energy dispersive
x-ray analysis. Certainly, in the field of carbon fibers, the use
of the SEM for structural studies is widespread. In particular,
fracture-face studies by SEM are invaluable in our understanding of
structure-property relationships.

Nevertheless, despite the simplicity of the technique, interpretation of the micrographs is not always straightforward. For
example, it is important to distinguish between a structure that is
the result of fracture and a structure that is inherent. Figure 8 is
a micrograph of a fracture face in a type I PAN-based carbon fiber.
The radiating structure is typical of the so-called hackle zone of
any brittle fracture but is not typical of the inherent structure.
Figure 9 is the fracture face of a type II PAN-based carbon fiber;
it is very similar in appearance to the fracture face of the type I
fiber. A cut bundle of mesophase pitch-based carbon fibers is shown
in Fig. 10. The sheetlike structures seen in this case are most
probably true structural features. By comparison, interpretation of
the normal surface of a carbon fiber might appear relatively simple;
it would be tempting to interpret the highly folded surface of the
type I carbon fiber in Fig. 8 as evidence of a fibrillar structure;

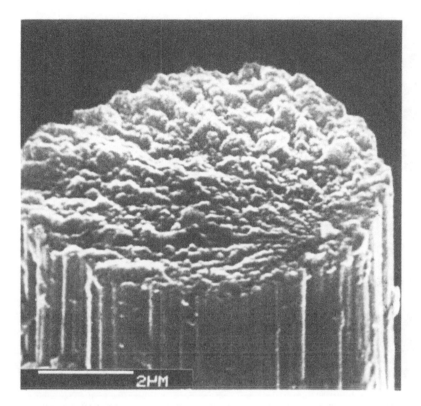

FIG. 8 SEM image of tensile fracture face from type I PAN-based carbon fiber.

as we shall see later, this is not the case. Evidently, it is important not to jump to conclusions when studying the results of SEM investigations.

B. Transmission Electron Microscopy

The transmission electron microscope (TEM) has been, and continues to be, another very important tool for investigating the structure of carbon fibers. Early work concentrated on thin fragments of material obtained by grinding or ultrasonic dispersion; more recent investigations have used longitudinal and transverse sections of fibers prepared by the difficult technique of ultramicrotomy.

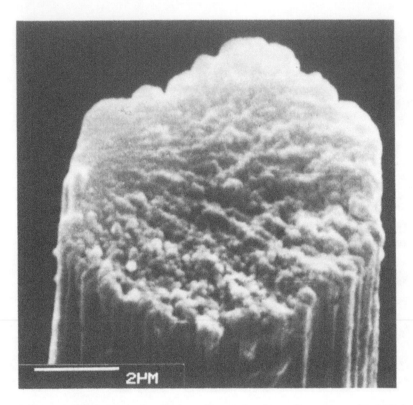

FIG. 9 SEM image of tensile fracture face from type II PAN-based
carbon fiber.

Operation of the TEM can utilize one of three imaging modes: bright-
field, dark-field, or lattice-fringe; the electron diffraction mode
is also useful.

The bright-field mode is the normal mode of operation, but the
dark-field mode is used increasingly in all carbon work since it
allows positive identification of the regions in a specimen that
contribute to a particular reflection in the diffraction pattern.
In dark-field mode, one of the diffraction spots is centered and all
other reflections are excluded by means of an aperture, so that the
image is formed only from those crystallites diffracting electrons
into the beam selected. Figure 11 is a typical dark-field (002)

FIG. 10   SEM image of cut face from mesophase pitch-based carbon fiber.

micrograph of a longitudinal section from a PAN-based type I carbon
fiber. The diffracting crystallites are seen as white regions on a
dark background; the edge of the structured region is, in fact, the
surface of the fiber.

In a TEM with a low spherical aberration coefficient (e.g.,
C_S = 1.6 mm), it is possible to obtain lattice-fringe images of the
0.34 nm turbostratic graphite layers at high magnification. If the
imaging is carried out with the undiffracted (zero-order) reflection
centered, the image is an "axial" or "multiple-beam" image; if the
zero-order beam and a 002 diffracted beam are tilted at equal angles
about the electron-optical axis of the microscope and all other
diffracted beams excluded, then we have a "tilted-beam" image.

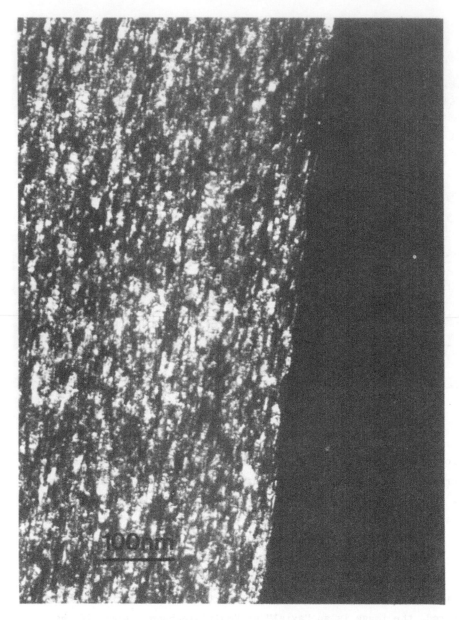

FIG. 11 TEM dark-field (002) image of a longitudinal section from
a type I PAN-based carbon fiber. (Reprinted with permission from
Ref. 54.)

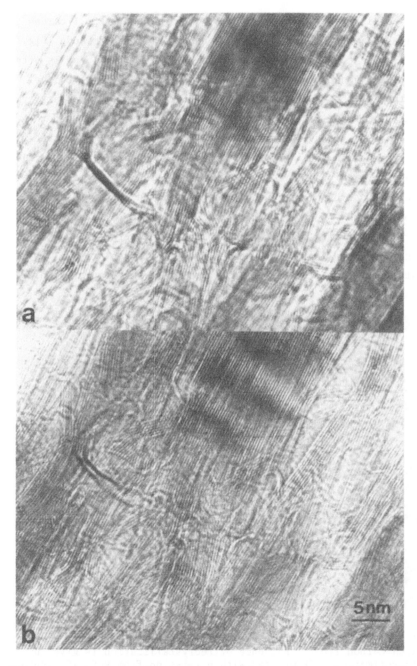

FIG. 12 High-resolution TEM lattice-fringe image showing 0.34 nm
layer planes in a type I PAN-based carbon fiber fragment: (a) axial
mode of imaging, (b) tilted-beam mode of imaging.

Figure 12 illustrates the same area of a fragment of type I carbon
fiber imaged in axial mode (Fig. 12a) and in tilted-beam mode (Fig.
12b). If the C_s of the microscope is high (say, 4.5 mm), as is often
the case when goniometer stages are in operation, then the tilted-
beam method may be the only possibility of imaging a carbon lattice.

The electron diffraction mode is particularly useful when the
longitudinal or transverse structure has zones of different structural organization. Using a "selected area" aperture it is possible
to select any region and record its selected area diffraction pattern.
This pattern can be digitized and the diffraction peaks analyzed by
analogous methods to those used in high-angle x-ray diffraction [11].
Figure 13 illustrates an electron diffraction pattern from a PAN-
based carbon fiber heat treated after nickel electroplating [15].
Recrystallization of the carbon has taken place with improved three-

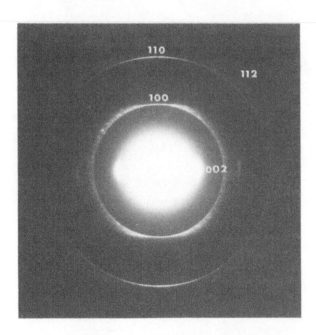

FIG. 13 Electron diffraction pattern from a nickel-treated type I
PAN-based carbon fiber. Evidence of recrystallization is given by
the 112 reflection. (Reprinted with permission from Ref. 15, copy-
right The Institute of Physics.)

dimensional order, as exemplified by the 112 reflection (arrow). This recrystallization leads to a greatly reduced tensile strength and modulus.

A complete account of high-resolution lattice-fringe imaging in TEM has been given by Millward and Jefferson [20]; an earlier account specific to PAN-based carbon fibers was given by Johnson and Crawford [21]. It must be emphasized that there is no exact one-to-one correspondence between layer planes in the specimen and lattice fringes in the image. This has been demonstrated with computer-simulated images [20] and can be observed both by changing the lattice image from multiple-beam mode to tilted-beam mode, as in Fig. 12, and by changing focus, as in Fig. 14 [21], where a crystallite (arrow) is seen after a number of equal increments of focus change. The lattice-fringe image, indeed any other high-resolution image, is formed by the transfer of information from the object to the image via the objective lens, a process affected by the spherical aberration coefficient and the level of focus of the objective lens. With a C_s of 1.6 mm or greater, the 0.34-nm graphite layer-plane spacing comes in a zone of the phase contrast transfer function (PCTF) where small changes in focus cause considerable change in the contrast of the lattice fringes. The best results in high-resolution TEM are obtained when the PCTF is as flat as possible (Scherzer focus); in other words, the maximum amount of information is transferred with the same contrast.

A simple test for the PCTF can be made with an optical diffractometer; then, any information gaps in image transfer are revealed as troughs between the peaks in the optical diffraction pattern. If, for example, gaps occur for spacings of the order of 0.5-0.7 nm, then there is no contrast in the image for such repeats. This can give a disordered carbon the appearance of a structure containing voids and is particularly important when interpreting the images of type II and type A PAN-based carbon fibers. Crawford and Marsh have demonstrated this effect and show how valid and invalid images can be distinguished [22]. Figure 15 demonstrates valid (Fig. 15a) and

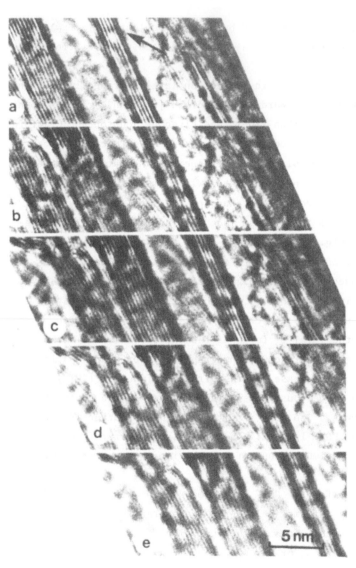

FIG. 14 A crystallite in a fragment of a type I PAN-based carbon
fiber imaged in axial mode. The contrast and the number of lattice
fringes vary with focus: (a) Δf = +162.5 nm, (b) Δf = +187.5 nm,
(c) Δf = +212.5 nm, (d) Δf = +237.5 nm, (e) Δf = +262.5 nm.
(Reprinted with permission from Ref. 21.)

invalid (Fig. 15b) images for a type A carbon fiber. Because of the
high layer-plane disorder, contrast in the valid image (Fig. 15a) is
low and it would appear that the "best" image is the invalid image
(Fig. 15b), in which contrast between crystallite and background is
enhanced by the "voids" introduced because of the information gap in
the PCTF, clearly seen in the optical transform of this image. The
optical transforms are inset and the fiber axis direction marked by
arrows.

IV. TENSILE PROPERTIES
A. Tensile Modulus

The theoretical tensile or Young's modulus of the straight-chain
segments of an organic polymer can be calculated from the force displacement curve of two adjacent atoms in the chain and the cross-
sectional area of the chain, to be around 200-300 GPa, dependent
upon the particular molecule. This establishes an insurmountable
limit to the modulus that can be achieved with long-chain molecules,
even when all the chains are unfolded. By comparison, the theoretical modulus of a carbon fiber composed of perfect layer planes is
very much higher, around 1000 GPa, a factor that first stimulated
the invention of carbon fibers [23]. With all carbon fibers, quite
independent of their origin, a well-established correlation exists
between preferred orientation and Young's modulus: the more highly
oriented the layer planes in the carbon fiber, the higher the tensile
modulus [6].

Two procedures that can give greatly increased tensile modulus
are hot stretching, which is better described as stress graphitization at high temperature [24] and can give moduli up to 650 GPa, and
boron doping [25], which can improve the tensile modulus to around
550 GPa. Unfortunately, these methods are not commercially viable,
but as they give values of modulus closer to the theoretical limit,
we may conclude that a high Young's modulus is not inordinately
difficult to attain, it is simply a matter of ensuring that the

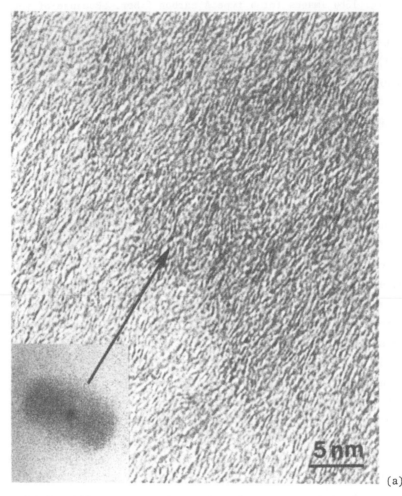

(a)

FIG. 15 High-resolution TEM lattice-fringe images from the
same area in a longitudinal section of a type A PAN-based
carbon fiber: (a) valid image with low contrast and flat

layer planes are aligned as perfectly as possible. Indeed, this
alignment has been easier to achieve with mesophase pitch-based
fibers, which are now available with a Young's modulus higher than
800 GPa (see Table 1). The enhanced crystallization brought about
by stress graphitization and boron doping is shown in Figure 16.
The crystallites are about 14 nm in width, considerably greater

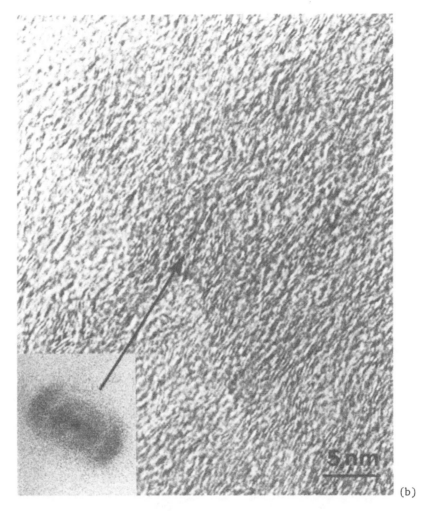

(b)

optical transform (inset), (b) invalid image with information gap in optical transform (inset). Fiber axis shown by
arrow.

than the width of a normal crystallite in a type I carbon fiber
(Table 2).

B. Tensile Strength

The theoretical tensile strength of a solid is more difficult to
evaluate than the theoretical tensile modulus [26]. One approach

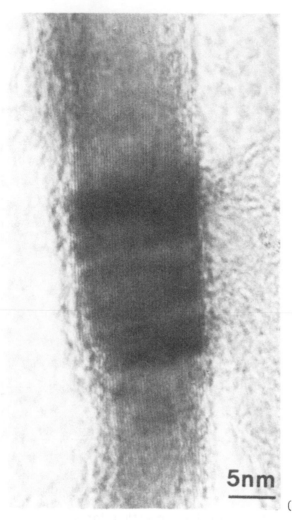

(a)

FIG. 16 Lattice-resolved crystallites of about 14 nm in width in
specially treated type I PAN-based carbon fibers: (a) stress
graphitized, (b) boron doped.

is the Orowan-Polanyi expression $\sigma_t = (E\gamma_a/a)^{1/2}$, which relates the
theoretical strength σ_t to Young's modulus E, the surface energy γ_a,
and the interplanar spacing a of the planes perpendicular to the
tensile axis. The ratio σ/E varies considerably but for many materials is in the range 0.1-0.2. For a perfect graphite the theoretical

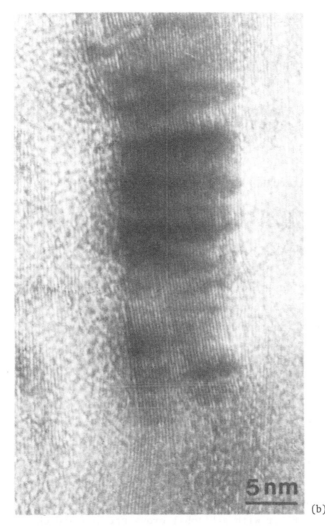

(b)

FIG. 16 (continued)

modulus is approximately 1000 GPa; consequently a theoretical strength
of 100 GPa might be expected. Because of defects, the practical
strength is always an order of magnitude lower than the theoretical
strength; thus in the most perfect form of fibrous carbon so far
produced, the graphite whisker, E is about 680 GPa and σ about 20 GPA
[27], a σ/E ratio of 0.03. The strongest carbon fibers have σ/E
ratios around 0.02. In brittle solids the defects are small cracks
that act as stress concentrators; they grow under the action of a

stress σ if they are greater than a critical size C as determined by the Griffith's relationship

$$\sigma^2 = \frac{2E\gamma_a}{\pi C}$$

If this relationship is applied to a graphite whisker, with $\gamma_a = 4.2$ J m^{-2}, the critical flaw size is 4.5 nm, and for a type I carbon fiber ($\sigma = 3.9$ GPa, E = 230 GPa) it is 40 nm.

C. Effect of Various Treatments

1. *Normal Heat Treatment*

According to theory, the tensile strength and tensile modulus should increase in proportion. However, from the earliest days of carbon fiber production, a decrease in tensile strength with increase in heat treatment temperature was reported for circular cross-section fibers [28]. For example, a carbon fiber produced at 1500°C had a strength of 2.33 GPa and a fiber produced at 2500°C had a strength of 1.8 GPa. This anomalous decline in strength stimulated considerable research on strength-structure relationships, much of it concerned with the possible presence of cracks and the known presence of pores or voids that, as discussed earlier, give rise to the prominent lobe-shaped small-angle x-ray diffraction pattern. It was thought intuitively that these cracks or voids might be responsible for the reduction in tensile strength at higher heat treatment temperatures.

LeMaistre and Diefendorf [29] noted that the decrease in strength did not occur in a carbon fiber with a bilobal cross section and characterized both circular and bilobal cross-section fibers as having circumferentially oriented layer planes. They proposed that Mrozowski cracks are generated in circular cross-section fibers during cooling from high temperatures because of the circumferential orientation of the layer planes. Although the layer planes in bilobal cross-section fibers are also circumferential, their predominant orientation along the long axis of the cross section was held to reduce the possibility of cleavage cracking.

2. Stress Graphitization

Johnson et al. [24], at Rolls-Royce, Ltd., showed that high-temperature stress graphitization of PAN-based carbon fibers reversed the strength decrease and reported that this was also the case with cellulose-based carbon fibers. The main structural changes were increases in L_c, the stacking size, and decreases in Z, the measure of preferred orientation. Johnson and Tyson [16] measured changes in the void diameter from analysis of the characteristic x-ray small-angle scattering and noted an increase from 0.9 nm for fibers heat treated at 1000°C to 2.9 nm for fibers heat treated at 2400°C. There was no significant change in the void diameter of stress-graphitized fibers from Rolls-Royce, although the increased crystallite size and perfection were evident when fragments were examined in the TEM. Figure 16a is a good example of a typical large crystallite. The size of the voids in the direction in which fracture proceeds, the transverse direction, is very much less than the critical flaw size and can be ruled out as contributing to strength limitation.

3. Neutron Irradiation and Boron Doping

Experiments carried out at the National Physical Laboratory [25] showed that a dose of neutrons sufficient to displace only 1 atom in 10,000 was capable of increasing both the modulus and strength of a PAN-based carbon fiber by 10%. Boron doping was found to increase Young's modulus by a similar amount, but tensile strength was unchanged. It was considered that neutron irradiation increased tensile strength as a consequence of a hardening mechanism whereby shearing of the layer planes was hindered by irradiation defects. Boron was assumed to induce conflicting effects; on the one hand, boron produces more perfect graphite, which would lower the shear strength and could thus be expected to reduce the tensile strength; on the other hand, boron acts as a solid-solution hardener, which might be expected to increase tensile strength; the net result was no change. The increased size and perfection of crystallites in boron-doped carbon fibers was immediately apparent on TEM investigation. Figure 17 gives a good impression of the enhanced crystallization as seen

FIG. 17 Dark-field micrograph of crystallites in boron-doped
type I PAN-based carbon fiber.

on a dark-field (002) micrograph. A typical large crystallite recorded with lattice resolution is illustrated in Fig. 16b.

D. Effect of Gross Defects

It has been known for some time, following the work of Moreton [30], that tensile strength measurements are gauge-length dependent and that the fibers produced earlier had a random distribution of flaws. Johnson [31] and Johnson and Thorne [32] investigated the fracture surfaces of carbon fibers by scanning electron microscopy. Tensile fracture was seen to occur at recognizable flaws located either at the surface or internally; they concluded that these flaws came from impurities originating in the PAN precursor. Sharp and Burnay [33] used a high-voltage TEM to investigate flaws; they found that there is no well-defined relationship between defect size and fracture strength; indeed, when defects occur in groups, fracture does not always occur at the largest defect. This result immediately raises doubt concerning predictions of tensile strength based on "weakest link" theories.

A series of experiments by Moreton and Watt [34] showed that when rigorous methods of dope filtration and clean-room spinning were applied in the production of PAN precursor fibers, carbon fibers could be produced that showed no decrease in tensile strength as heat treatment temperature and modulus increased. When samples were deliberately contaminated at the final stage of the spinning process with carbon black, silica, and ferric oxide, the individual fiber strengths were lower and the coefficients of variation were greater than for the clean samples. A more recent analysis by Jones et al. [35] of flaws in high-strength carbon fibers from a contaminated mesophase pitch precursor used an SEM with an energy dispersive x-ray analyzer attached. Fracture surfaces were examined after tensile failure and were always found to contain recognizable defects. The regions around these defects exhibited traces of contaminants, such as iron, nickel, and chromium. Fibers prepared under clean conditions failed at higher strengths, for example, 3.0-4.0 GPa, compared with 1.5-2.5

GPa for the contaminated specimens; in addition, it was usually difficult to detect defects in the fracture faces.

It is now widely recognized that defects arising from contaminating particles are the major cause of reduced strength at higher heat treatment temperatures and that to evaluate the intrinsic strength of carbon fibers requires a complete understanding of their macro- and microstructural organization and the behavior of these structures under tensile deformation.

V.   FIBER STRUCTURE

A.   Macrostructure

Polished cross sections of carbon fibers from various sources, when examined in an optical microscope with polarized light, exhibit "Maltese cross" patterns with interference colors in alternate quadrants. Butler and Diefendorf [36] interpreted the patterns from PAN-based carbon fibers in terms of a sheath-core heterogeneity, the outer sheath having turbostratic graphite crystallites that are larger and more highly oriented than in the core. Watt and Johnson [37] examined polished longitudinal sections by optical methods and showed that an outer sheath can originate from fully stabilized material and an inner core from partially oxidized PAN. Further optical studies [38-40] suggested a circumferential crystallite orientation in fully stabilized fibers of circular cross section and a radial orientation in understabilized material. Fibers having a bilobal cross section appeared to have an outer sheath with crystallites oriented parallel to the surface and a core with random crystallite orientation.

Johnson [41] has raised the possibility that the optical anisotropy seen in carbon fibers is not caused by structural inhomogeneity at all. Evidence was presented that dark-field TEM images of longitudinal sections show no sheath-core differentiation, and that, in transverse sections, neither lattice-fringe images nor electron diffraction patterns show anything other than a random arrangement of layer planes in any part of the section. Johnson concludes that optical anisotropy is caused not by structural heterogeneity but by strain birefringence, stress in the fibers being caused by differential shrinkage.

Other evidence purporting to support macrostructural organization was obtained by Barnet and Noor [42], who examined circular-section PAN-based carbon fibers in an SEM after ion etching; they interpreted their images in terms of a highly crystalline, circumferentially organized sheath surrounding a radially structured core comprising both crystallites and large voids. It would seem that this structure is more the result of the etching process than a fundamental mode of organization. The typical outcome of macrostructural studies is a schematic diagram similar to that of Fig. 18, which portrays a circumferential sheath-radial core model.

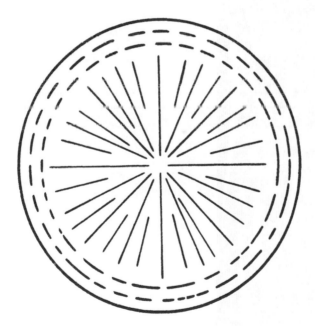

FIG. 18 Schematic diagram of structure in type I PAN-based carbon fiber depicting circumferential sheath and radial core. (Reprinted with permission from Ref. 46.)

B. Microstructure

Although some of the earliest TEM studies of fragments of carbon
fibers were held to indicate the presence of a fibrillar structure,
subsequent studies of high-resolution lattice-fringe images, for
example, Fig. 12, have shown this not to be the case. For cellulose-
based fibers, two groups [16,43] have suggested a "basket-weave"
model of layer planes, whereas a third group [44] suggested a ribbon-
like model in which curvilinear layer planes are packed side by side
enclosing voids of an approximately needle shape (Fig. 19). This
model was later proposed for PAN-based carbon fibers [6] and has
some merit for relating Young's modulus and preferred orientation.
A more realistic model of the structure of a type I PAN-based carbon

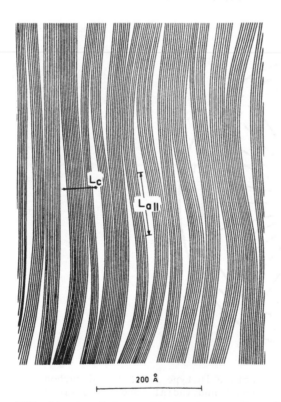

FIG. 19 Curvilinear layer-plane model of structure in type I PAN-
based carbon fibers. (Reprinted with permission from Ref. 6.)

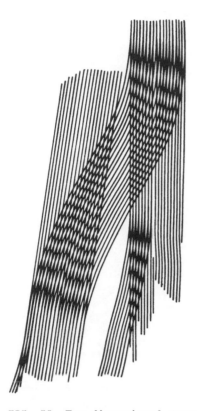

FIG. 20 Two-dimensional representation of the interlinking of layer planes in the longitudinal direction of type I PAN-based carbon fibers.

fiber, embodying the results of many observations [45] on both fragments and longitudinal sections of carbon fibers, is given in Fig. 20 in the form of a simple two-dimensional representation of the complex three-dimensional interlinking of layer planes forming crystallites that enclose sharp-edged voids.

Later work by Bennett and Johnson [46] showed that some type I fibers can have skin-core heterogeneity. This skin, which was never more than 0.5 μm thick, should not be confused with the very much thicker sheath referred to earlier, which may be considered an artefact except in those fibers that are not fully stabilized. Both dark-field and lattice-fringe images and electron diffraction patterns, showed that the skin had larger and better oriented

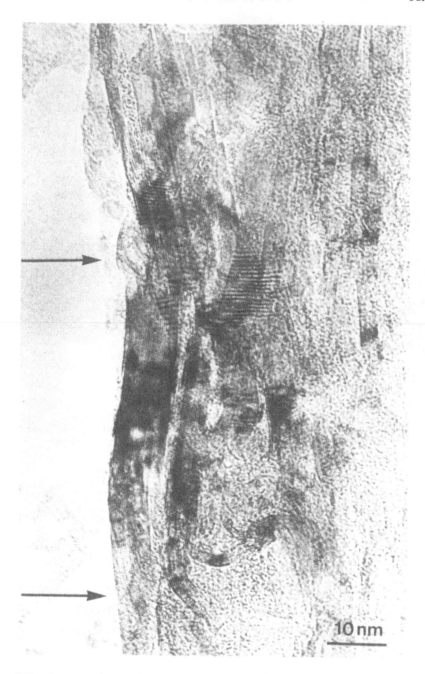

FIG. 21 Lattice-fringe image from a longitudinal section of a type I
PAN-based carbon fiber depicting a skin region. The fiber surface is
indicated by arrows.

crystallites than the core. Figure 21 is a typical example of the
skin region of the fiber seen in longitudinal section; the surface
is indicated by arrows. Characterization parameters for the skin
and core regions [7] were obtained by the application of electron
diffraction techniques [11] analogous to the x-ray diffraction
methods used for whole fibers.

Lattice-fringe images of transverse sections reveal a very
complicated situation. In the skin region, the layer planes are
essentially parallel to the surface, but in addition to the type of
complex crystallite interlinking seen in longitudinal section, many
layer planes fold through angles up to 180° in "hairpin" fashion.
This is particularly noticeable in Fig. 22, which is from a type I

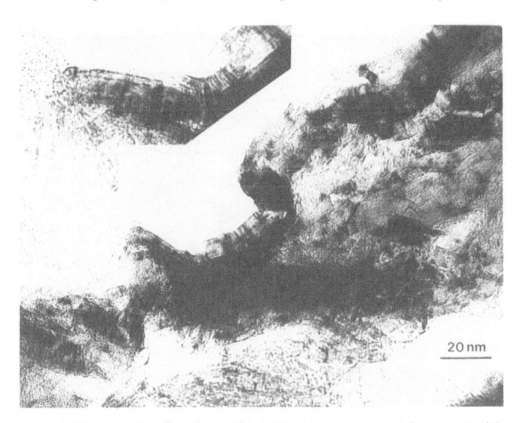

FIG. 22 Lattice-fringe image from a transverse section of a type I PAN-
based carbon fiber with a thin skin. Inset is a detail at higher magnification. (Reprinted with permission from Ref. 46.)

PAN-based carbon fiber in cross section. A typical region of the
core is illustrated in Fig. 23. In the lateral direction throughout
the core, the layer planes are folded extensively, providing coher-
ence over large cross-sectional areas of the fiber. These micrographs provide conclusive evidence that PAN-based carbon fibers do
not have a fibrillar structure and do not have a radially oriented
core. Primarily, the structural organization in cross section is
random. A convenient model of structure based on these observations
of transverse sections is shown in Fig. 24.

At the same time [46], detailed TEM studies of type II and type A
fibers showed that there is no skin-core heterogeneity and that the
smaller crystallites seen in these fibers have much less well ordered

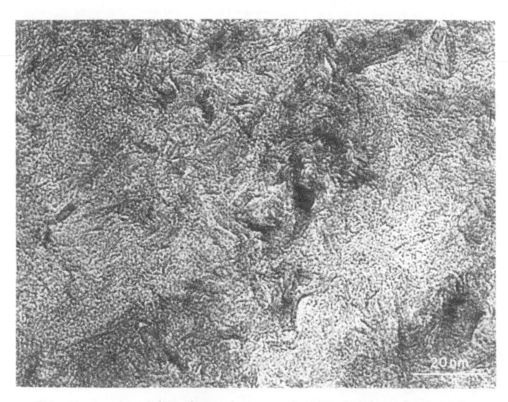

FIG. 23 Lattice-fringe image from a transverse section of a type I
PAN-based carbon fiber showing a typical core region.

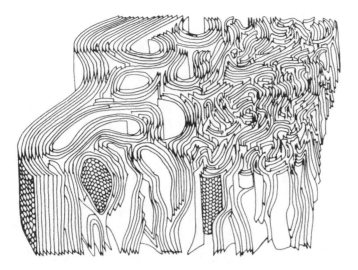

FIG. 24 Schematic three-dimensional model of structure in a type I PAN-based carbon fiber. The layer planes are highly interlinked in both longitudinal and transverse directions.

layer planes. A typical lattice-fringe image of a type II PAN-based carbon fiber is shown in Fig. 25. The structure of a type A fiber was illustrated in Fig. 15. If disordered layer planes are substituted into Figs. 30 and 24, then we have reasonable models of structure in type II fibers. If extremely disordered layer planes are substituted, then we have models of type A fibers.

After detailed dark-field and high-resolution TEM studies of two type I PAN-based carbon fibers (Rigilor AG and Celion GY-70),[*] Oberlin [47] has produced an alternative model in which the layer planes have a variable transverse radius of curvature that decreases continuously from the surface to the center, thus giving the appearance of skin-core heterogeneity. Oberlin [48] also maintains that the tensile strength of type I fibers is a function of the radius of curvature of the layer planes. The tensile strength of other types of PAN-based carbon fibers is said to depend on a so-called compactness index.

---

[*]Rigilor is the registered trademark of Serofim, a subsidiary of Rhone-Poulenc-Textile and LeCarbone-Lorraine. Celion is the registered trademark of Celanese Corporation.

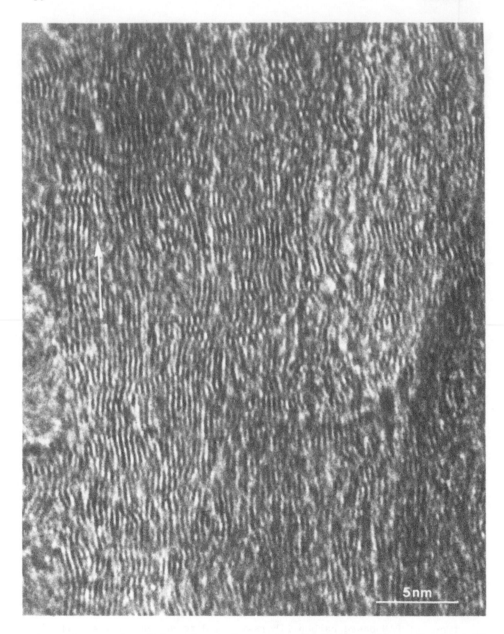

FIG. 25   Lattice-fringe image from a longitudinal section of a type II
PAN-based carbon fiber.

VI. FRACTURE MECHANISMS

A. Tensile Failure

Difficulties in describing fracture mechanisms in terms of dislocation pileup at grain boundaries, the unbending of curved ribbons, the presence of density fluctuations, or yield processes involving local shear deformation and slippage were discussed in the review by Reynolds [1]. Although a wide range of internal defects was observed, no simple relationship could be found among flaw diameter, fiber strength, and surface free energy. This led to the proposal by Reynolds and Sharp [49] of a crystallite shear limit for fiber fracture.

This mechanism of fracture is based on the idea that crystallites are weakest in shear on the basal planes. When tensile stress is applied to misoriented crystallites locked into the fiber structure, the shear stress cannot be relieved by cracking or yielding between basal planes. The shear strain energy may be sufficient to produce basal plane rupture in the misoriented crystallite and, hence, a crack that will propagate both across the basal plane and, by transference of shear stress, through adjacent layer planes. A schematic diagram of a misoriented crystallite well locked into the surrounding crystallites is shown in Fig. 26a. When stress is applied, basal plane rupture takes place (Fig. 26b) and proceeds throughout the local region (Fig. 26c). However, before a crack can propagate through a fiber and cause failure, either one of two conditions must be fulfilled.

1. The crystallite size in one of the directions of propagation of a crack, that is, either L_c or $L_{a\perp}$, must be greater than the critical flaw size C for failure in tension (C is about 40 nm for a type I fiber).

2. The crystallite that initiates catastrophic failure must be sufficiently continuous with its neighboring crystallites for the crack to propagate.

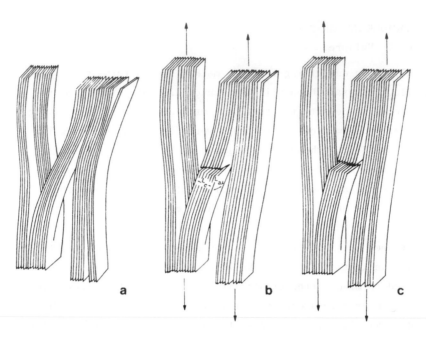

FIG. 26 Reynolds-Sharp mechanism of tensile failure. (a) Misoriented
crystallite linking two crystallites parallel to fiber axis. (b) Tensile stress exerted parallel to fiber axis causes layer-plane rupture
in direction $L_{a\perp}$; crack develops along $L_{a\perp}$ and L_c. (c) Further exer-
tion of stress causes complete failure of misoriented crystallite.
Catastrophic failure occurs if the crack size is greater than the
critical size in either the L_c or $L_{a\perp}$ directions. (Reprinted with
the permission of Chapman and Hall from Ref. 50.)

The first condition is not normally fulfilled because both L_c and $L_{a\perp}$
are much less than C, as in Fig. 26, although the effective value of
$L_{a\perp}$ is considerably greater than the values measured by x-ray diffraction, in which layer planes are curved or hairpin shaped (see Fig. 23).
The second condition is most likely to be satisfied in those regions
of enhanced crystallization and misorientation that have been observed
around a defect [33,46]. The fortuitous continuity of large crystallites at a large angle of misorientation may well explain why a fiber
can fail at the smaller of two defects. Essentially, it is the presence of large misoriented crystallites that cause failure, not the
presence of a hole.

In a recent study of tensile failure [50], specimens from an old batch of type I PAN-based carbon fibers containing many flaws were stressed to failure in glycerol. This enabled the fracture ends to be preserved intact for subsequent examination, first by SEM, and then, after embedding and sectioning, by TEM. Internal flaws that did not initiate failure were seen to have walls containing crystallites arranged mainly parallel to the fiber axis, as illustrated by the dark-field (002) micrograph of Fig. 27. Internal and surface flaws that did initiate failure often showed evidence of large misoriented crystallites in the walls of the flaws. Continuity of crystallites in the walls may well give rise to values of $L_{a\perp}$ that exceed the critical flaw size. The failure-initiating flaw in the fracture face of Fig. 28 has a very extensive well-graphitized wall; evidently the critical size was exceeded in this case.

Further proof for the concept that large misoriented crystallites, together with continuity of structure in the walls of flaws, cause fiber failure under stress, is found in an earlier study of lignin-based carbon fibers [51]. These fibers, which had very inferior tensile strengths, contained many flaws in the form of inclusions caused by catalytic graphitization around impurity particles in the precursor material. A typical bright-field image of one such inclusion is shown in Fig. 29. These inclusions are solids of revolution whose walls are often contiguous with the normal layer-plane structure of the fiber. When stressed, there must be misoriented crystallites at all angles under tension; consequently, layer-plane failure takes place by the Reynolds-Sharp mechanism, and a crack is initiated that is able to propagate around the wall of the inclusion, thus precipitating fiber failure.

It is of interest to note that, despite their overall better preferred orientation, the skin regions of type I PAN-based carbon fibers were usually found to contain several large misoriented crystallites. Thus, in the absence of gross defects, the skin itself is a strength-limiting feature [46]. Figure 30 is a good example of a flaw in the skin of a type I fiber. It comprises a protrusion at

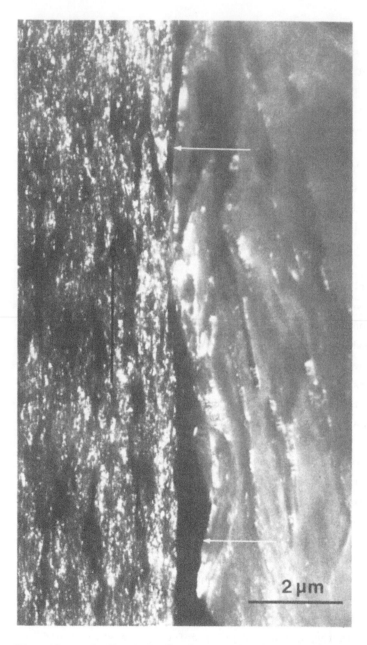

FIG. 27 Dark-field (002) micrograph of longitudinal section through
a flaw near a fracture surface. This flaw, whose edge is marked with
white arrows, did not initiate failure, and its wall contains crystallites arranged mainly parallel to the fiber axis (black arrow).
(Reprinted with the permission of Chapman and Hall from Ref. 50.)

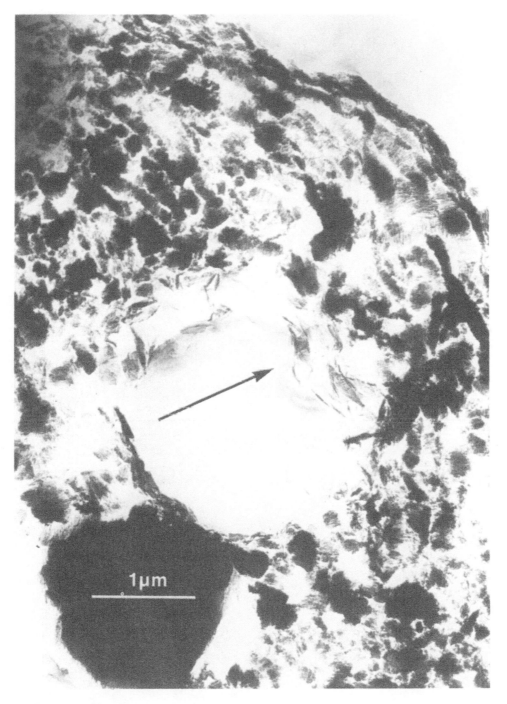

FIG. 28 Bright-field micrograph of a fracture face in transverse
section. Failure was initiated by the large internal flaw, the walls
of which contain extensive well-graphitized sheets (arrowed).
(Reprinted with the permission of Chapman and Hall from Ref. 50.)

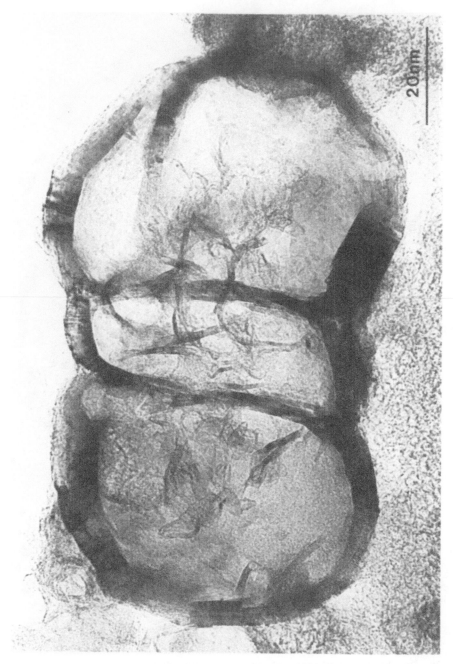

FIG. 29 Bright-field micrograph of a typical inclusion in a lignin-based carbon fiber.

FIG. 30 Lattice-fringe image of hairpin flaw in skin of type I
PAN-based carbon fiber seen in longitudinal section. Crystallites
run at 45° to fiber axis (arrow) through skin and into core.
(Reprinted with the permission of Chapman and Hall from Ref. 46.)

the surface consisting of layer planes bent through a hairpin and
entering the core of the fiber at an angle of roughly 45°. Again,
one can visualize the layer planes under shear stress failing by the
Reynolds-Sharp mechanism and initiating total failure of the fiber.

Structural studies of recent high-modulus and high-strength
PAN-based carbon fibers show no evidence of a skin; furthermore,
fracture faces do not normally show any recognizable initiating
defect, such as an internal or surface void. A dark-field TEM image
of a longitudinal section through a recent type I fiber is shown in
Fig. 11; the crystal structure is seen to be homogeneous throughout,
there is no indication of a skin, and there is no indication of a
gross macroscopic defect in the mirror zone of the fracture face
(Fig. 8). Evidently, improved structural homogeneity and the absence
of flaws leads to improved tensile strength.

B. Flexural Failure

Although considerable attention has rightly been paid to the tensile
deformation of carbon fibers, much less attention has been focused
on deformations that involve the compression or combined compression
and tension of a fiber. The reduction of tensile strength by the
insertion of a knot or loop is a measure of brittleness; the size of
the knot or loop at failure can be considered a measure of flexibility. These two parameters should be of interest to those concerned with handling both carbon fibers and carbon fiber composites
and with the behavior of the composites under flexural deformation.
In addition, Thorne [52] suggests that loop tests for a carbon fiber
can give a value for the effective tensile strength at a gauge length
as low as 0.1 mm. However, this suggestion is based on a rather over-
simplified estimate of the strain at the surface of a fiber constrained
in a loop, leading to excessively high values.

DaSilva and Johnson [53] have caused a number of commercially
available carbon fibers to be stressed to failure under both tensile
and flexural deformation (knot test). These tests were carried out
both in air and in glycerol. In air, quantitative measurements were

TABLE 3　Flexural Properties of Carbon Fibers[a]

Carbon fiber	E (GPa)	σ (GPa)	d (μm)	K (MPa)	σ/K (×1000)	Rad. (μm)
Type I						
HM-S Grafil	340	2.10	7.0	63	0.33	370
Thornel P (VSB-32-0)	380	1.90	10.0	10	1.90	840
Celion GY-70	539	1.96	~8.4	6	3.27	980
Type II						
Sta-grade Besfight	240	3.73	7.0	93	0.40	230
Thornel 300	239	3.11	7.0	62	0.47	290
Type A						
XA-S Grafil	230	2.90	7.0	34	0.91	300

[a]E, Young's modulus; σ, tensile strength; d, fiber diameter, approximate average for bilobal fiber; K, knot strength; σ/K, brittleness; Rad., knot diameter before failure.

made; in glycerol, fracture ends were preserved for subsequent examination by SEM. Fiber samples containing knots were also stressed to failure on the stage of an SEM and the diameter of the knot just before failure measured from micrographs.

Results for a number of different carbon fibers are given in Table 3 [54], where it is seen that among the type I fibers, the mesophase pitch-based fibers (Thornel P; VSB-32-0)[*] and the PAN-based Celion GY-70 fibers are very much more brittle and inflexible than the PAN-based HM-S Grafil fibers.[†] Among the type II and type A fibers, the Besfight fiber is least brittle and most flexible, and has the highest tensile and knot strength.[‡] It is of some interest to note that this fiber has the most perfectly circular cross section of all the fibers studied (see Fig. 31, which illustrates a fracture face after tensile failure). This investigation revealed very little evidence for fracture-initiating flaws in any of the fracture faces examined after tensile failure; this suggests that the population of defects in all these fibers is low. The reason for the marked

---

[*]Thornel is the registered trademark of Union Carbide Limited.
[†]Grafil is the registered trademark of Courtaulds Limited.
[‡]Besfight is the registered trademark of Toho Beslon Company Limited.

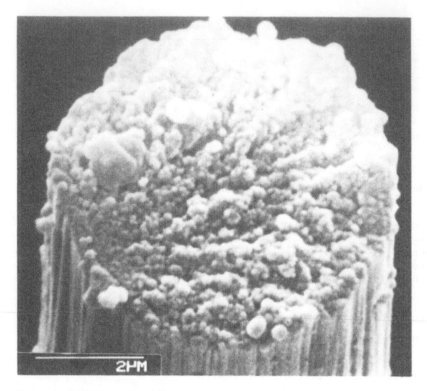

FIG. 31   SEM image of fracture face after tensile failure of Besfight
fiber (type II).   (Reprinted with the permission of Chapman and Hall
from Ref. 53.)

differences in brittleness and flexibility must be sought in the fine
structure of the fibers.

The fracture faces of circular PAN-based carbon fibers after
flexural failure are distinctly different from the faces after tensile
failure. Figure 32 shows the fracture face from a Besfight fiber
after flexural failure in glycerol; this and most other fracture faces
from type II fibers exhibit a rough striated area and a relatively
smooth area corresponding to the regions of the fiber under tension
and compression, respectively. The type I PAN-based fiber HM-Grafil
(Fig. 33) has a much more deeply corrugated area between the smooth
compression zone and the rough tension zone; this probably corresponds
to a region of the fiber under maximum shear stress.

FIG. 32   SEM image of fracture face after flexural failure of
Besfight fiber (type II).   (Reprinted with the permission of
Chapman and Hall from Ref. 53.)

The fracture surface of the Thornel mesophase pitch-based fiber
is remarkably similar in flexural failure (Fig. 34) to the surface
produced by tensile failure (Fig. 35) and to faces produced simply
by cutting the fiber (Fig. 10). Evidently, sheetlike features are
real structures within these fibers, which, although not as radial
in structure as some of the earlier mesophase pitch fibers, are still
predominantly sheetlike. These sheets can be expected to propagate
a transverse crack, according to the Reynolds-Sharp fracture mechanism, much better than a random structure, because of the large effective crystallite size in the direction of the sheet, which may well
exceed the critical flaw size. The uniformity of structure seen in
a knot fracture face may indicate the simultaneous failure in tension

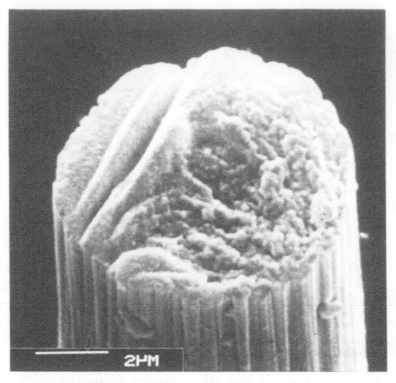

FIG. 33 SEM image of fracture face after flexural failure of
HM-Grafil fiber (type I). (Reprinted with the permission of
Chapman and Hall from Ref. 53.)

of sheets through that part of the section. When the fibers are
deformed into a knot, it is inevitable that many more crystallites
are misoriented; consequently, the chances of layer-plane rupture
and subsequent fracture are increased.

Possibly the first SEM images of fracture faces in high-modulus
carbon fibers were those reported by Jones and Johnson [55], who
also demonstrated evidence for buckling in the compressive region
during bending in a loop test. Buckling failure was considered to
take place before ultimate failure from the region under tension.
We have considerable experience of this type of failure in Kevlar-
type fibers under flexural deformation [56]. Despite very careful

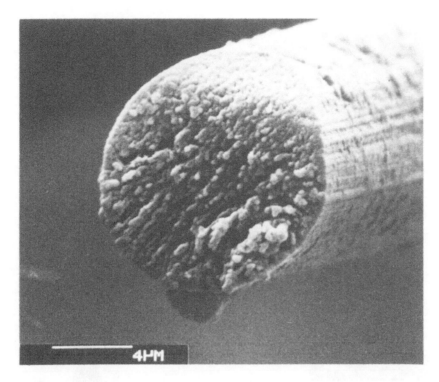

FIG. 34   SEM image of fracture face after flexural failure of Thornel
mesophase pitch-based fiber.   (Reprinted with the permission of
Chapman and Hall from Ref. 53.)

observation of all fibers during knot closure in the SEM, and despite
careful study of all the micrographs recorded before fiber failure,
no evidence of buckling was noted in the carbon fibers that were the
subject of the investigation reported by DaSilva and Johnson [53].

Ewins and Potter [57] have suggested that a Reynolds-Sharp type
of mechanism, common to both tension and compression regions, is
responsible for the initiation of failure in fibers fractured within
a composite during longitudinal compression.   Simultaneous failure
in tension and compression regions may well be the reason that buckling has not been observed in these knot tests.   Whether there is
any internal buckling not noticeable in SEM is a matter for further
study.

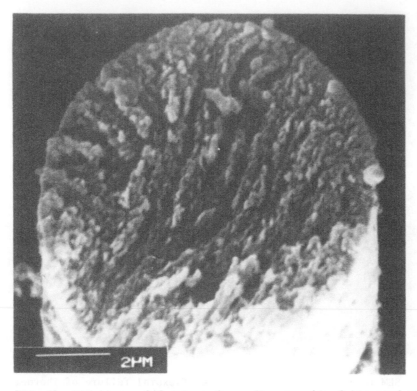

FIG. 35 SEM image of fracture face after tensile failure of Thornel mesophase pitch-based fiber. (Reprinted with the permission of Chapman and Hall from Ref. 53.)

VII. PREDICTIONS OF ULTIMATE TENSILE STRENGTH

Reynolds and Sharp [49] showed that the crystallite shear strain to failure ε_s is given by

$$\varepsilon_s = \frac{\varepsilon E}{c_{44}} \sin \phi_c \cos \phi_c$$

where ε is the strain to failure, E the Young's modulus, ϕ_c the local crystallite misorientation, and c_{44} is the shear modulus. Using a practical value of 20% for ε_s and the criterion that failure may occur when a crystallite is misoriented at an angle twice the average misorientation, as determined by x-ray diffraction, a type I carbon fiber with mean $\phi_c = 5.0°$ and E = 400 GPa will have a tensile strain to failure of 1.3% and tensile strength of 5.2 GPa. From the Griffith's equation given earlier, the critical flaw size C is about 50 nm. A crack produced by basal plane rupture will propagate in the directions measured by x-ray diffraction as $L_{a\perp}$ and L_c; these are normally an order of magnitude less than C in a type I fiber. However, continuity of structure in the neighborhood of a defect may lead to C being exceeded and failure taking place.

If all defects are eliminated, what is the upper limit to tensile strength in terms of crack propagation in a single crystallite? The largest crystallite measured in a study of crystallite size distribution [58] had a stacking size of L_c of 18 nm; crystallites of about 20 nm extent in the direction of L_a can be seen in transverse sections. If we take 20 nm as a reasonable value for the critical flaw size, then we have an ultimate tensile strength of about 7.5 GPa for a Young's modulus of 400 GPa; this implies a strain to failure of 1.9% and a misorientation angle of 6.5°.

The chance of larger crystallites in type I fibers and the greater chance of structural continuity mean that an estimate of around 5 GPa is probably a reasonable practical upper limit to tensile strength for type I fibers. However, the lower probability of finding large misoriented crystallites in type II or type A fibers suggests that an upper limit of 7.5 GPa is not unreasonable in terms of the Reynolds and Sharp mechanism, if this is applicable to these types of fiber.

Although a fiber does not always fail at the largest flaw, so-called weakest link theories have been used to predict intrinsic carbon fiber strength since the gauge-length dependence was first reported. Chwastiak et al. [59] carried out a Weibull statistical analysis on mesophase pitch-based carbon fibers and predicted a strength of 5.3 GPa at 0.3-mm gauge length. In the recent detailed paper on flaws in pitch-based fibers by Jones et al. [35], which demonstrated that careful spinning can eliminate the larger flaws, it was suggested that surface pits then become a limiting factor, and the predicted tensile strength at 0.3-mm gauge length is reduced to 3.8 GPa. Beetz [60,61] has carried out a detailed Weibul analysis

using the bimodal probability of failure as found with mesophase
pitch-based fibers; this leads to a predicted strength of around
3.5 GPa at 0.3-mm gauge length. The gauge length of 0.3 mm is a
good estimate of the load transfer length in composites.

The weakest link approach to the calculation of intrinsic
strength is essentially empirical in that it involves an extrapola-
tion from data on defective material. In the absence of defects it
might be thought that no useful predictions can be made; certainly,
several PAN-based carbon fibers are now available with a tensile
strength greater than 4.0 GPa (see Table 1). It would be interesting
to find the extrapolations from these newer, relatively flaw free
fibers. The weakest link approach probably gives too low a value to
the ultimate tensile strength; the structural limitations based on
the Reynolds-Sharp mechanism of failure probably give a higher and
more realistic value.

VIII. CONCLUSION

It would seem that we can now identify the requirements for optimum
physical properties in a PAN-based carbon fiber, or indeed in a carbon
fiber of any origin. The perfect carbon fiber should be as nearly
circular as possible, without skin-core heterogeneity, and without
gross defects arising from catalytic graphitization by contaminant
particles in the precursor; it must also have high preferred orientation with a random structure of well-interlinked and folded layer
planes, rather than a structure of radial or otherwise organized
sheets. Type I fibers may well be limited to an ultimate tensile
strength, at short gauge length, of around 5.0 GPa, by their structural composition of large crystallites, some of which will be misoriented sufficiently to cause failure by the Reynolds-Sharp mechanism
of fracture. The smaller crystallites of the type II and type A
fibers should allow a higher limit to tensile strength, possibly
around 7.5 GPa. It appears that the random structure of PAN-based
carbon fibers leads to more flexibility and less brittleness than is
the case with the more sheetlike structures of the mesophase pitch-

based fibers, although it must not be overlooked that this sheetlike structure can confer very desirable properties in terms of much increased thermal and electrical conductivity, which are required for certain end uses.

ACKNOWLEDGMENTS

I wish to thank my former research students and assistants, C. N. Tyson, the late C. Oates, D. Crawford, S. C. Bennett, B. P. Saville, and K. J. Masters for their invaluable work on structural characterization, and J. L. G. DaSilva for his work on brittleness and flexibility. I am deeply indebted to W. Johnson and the late W. Watt F.R.S., both formerly of the Royal Aircraft Establishment Farnborough, for collaboration and discussion over a long period, and to the Ministry of Defence, Procurement Executive, for technical and financial assistance. I would also like to thank Celanese Corporation, Courtaulds Ltd., Toho Beslon Company Ltd., and Union Carbide Ltd., for the provision of samples.

REFERENCES

1. W. N. Reynolds, in *Chemistry and Physics of Carbon*, Vol. 11 (P. L. Walker and P. A. Thrower, eds.), Marcel Dekker, New York, 1973, p. 1.

2. L. S. Singer, U.S. Patent 4,005,183 (January 25, 1977).

3. L. S. Singer, *Carbon 16*, 409 (1978).

4. R. Bacon, in *Chemistry and Physics of Carbon*, Vol. 9, (P. L. Walker and P. A. Thrower, eds.), Marcel Dekker, New York, 1973, p. 1.

5. W. Ruland, in *Chemistry and Physics of Carbon*, Vol. 4 (P. L. Walker, ed.), Marcel Dekker, New York, 1968, p. 1.

6. A. Fourdeux, R. Perret, and W. Ruland, *Proceedings of the First International Conference on Carbon Fibres*, Plastics Institute, London, 1971, p. 57.

7. D. J. Johnson, *Phil. Trans. R. Soc. Lond. A294*, 443 (1980).

8. W. Ruland and H. Tompa, *J. Appl. Crystallogr. 5*, 225 (1972).

9. W. Ruland and H. Tompa, *Acta Crystallogr. A24,* 93 (1968).

10. A. M. Hindeleh, D. J. Johnson, and P. E. Montague, in *Fiber Diffraction Methods, ACS Symposium No. 141* (A. D. French and K. H. Gardner, eds.), American Chemical Society, Washington, D.C., 1983, p. 149.

11. S. C. Bennett, D. J. Johnson, and P. E. Montague, *Proceedings of the Fourth London International Conference on Carbon and Graphite 1974,* Society of Chemical Industry, London, 1976, p. 503.

12. W. O. Statton, J. L. Koenig, and M. Hannon, *J. Appl. Phys. 41,* 4290 (1970).

13. R. D. B. Fraser, T. P. MacRae, and A. Miller, *Nature 203,* 1231 (1964).

14. A. M. Hindeleh and D. J. Johnson, *Polymer 13,* 423 (1972).

15. D. J. Johnson and C. N. Tyson, *J. Phys. D, Appl. Phys. 2,* 787 (1969).

16. D. J. Johnson and C. N. Tyson, *J. Phys. D, Appl. Phys. 3,* 526 (1970).

17. R. Perret and W. Ruland, *J. Appl. Crystallogr. 1,* 308 (1968).

18. R. Perret and W. Ruland, *J. Appl. Crystallogr. 2,* 209 (1969).

19. I. Tomizuka and D. J. Johnson, *Yogyo-Kyokai-Shi 86,* 42 (1978).

20. G. R. Millward and D. A. Jefferson, in *Chemistry and Physics of Carbon,* Vol. 14 (P. L. Walker and P. A. Thrower, eds.), Marcel Dekker, New York, 1976, p. 1.

21. D. Crawford and D. J. Johnson, *J. Microsc. 94,* 51 (1971).

22. D. Crawford and H. Marsh, *J. Microsc. 109,* 145 (1977).

23. W. Watt, L. N. Phillips, and W. Johnson, *Engineer 221,* 815 (1969).

24. J. W. Johnson, J. R. Marjoram, and P. G. Rose, *Nature, 221,* 357 (1969).

25. S. Allen, G. A. Cooper, D. J. Johnson, and R. M. Mayer, *Proceedings of the Third London International Conference on Carbon and Graphite 1970,* Society of Chemical Industry, London, 1970, p. 456.

26. N. H. Macmillan, *J. Mater. Sci. 7,* 239 (1972).

27. G. Dorey, *Phys. Technol. 11,* 56 (1980).

28. R. Moreton, W. Watt, and W. Johnson, *Nature 213,* 690 (1967).

29. C. W. LeMaistre and R. J. Diefendorf, *Tenth Biennial Conference on Carbon,* Bethlehem, American Carbon Society and Lehigh University, 1971, p. 163.

30. R. Moreton, *Fibre Sci. Technol. 1,* 273 (1968).

31. J. W. Johnson, *Appl. Polym. Symp. 9,* 229 (1969).

32. J. W. Johnson and D. J. Thorne, *Carbon 7,* 659 (1969).

33. J. V. Sharp and S. G. Burnay, *Proceedings of the First International Conference on Carbon Fibres,* Plastics Institute, London, 1971, p. 68.

34. R. Moreton and W. Watt, *Nature 247,* 360 (1974).

35. J. B. Jones, J. B. Barr, and R. E. Smith, *J. Mater. Sci. 15,* 2455 (1980).

36. B. L. Butler and R. J. Diefendorf, *Ninth Biennial Conference on Carbon,* Boston College, American Carbon Society and Boston College, 1969, p. 161.

37. W. Watt and W. Johnson, *Proceedings of the Third London International Conference on Carbon and Graphite,* London, Society of Chemical Industry, 1970.

38. R. H. Knibbs, *J. Microsc. 94,* 273 (1971).

39. B. J. Wicks and R. A. Coyle, *J. Mater. Sci. 11,* 376 (1976).

40. R. J. Diefendorf and E. W. Tokarsky, *Polym. Eng. Sci. 15,* 150 (1975).

41. W. Johnson, *Nature 279,* 142 (1979).

42. F. R. Barnet and M. K. Noor, *Carbon 11,* 281 (1973).

43. J. A. Hugo, V. A. Phillips, and B. W. Roberts, *Nature 226,* 144 (1970).

44. A. Fourdeux, R. Perret, and W. Ruland, *C. R. Ser. C. 269,* 1597 (1969).

45. D. Crawford, Ph.D. Thesis, University of Leeds, England, 1972.

46. S. C. Bennett and D. J. Johnson, *Carbon 17,* 25 (1979).

47. A. Oberlin, *J. Microsc. Spectrosc. Electron 7,* 327 (1982).

48. M. Guigon, A. Oberlin, and G. Desarmot, *Proceedings of the Sixth London International Conference on Carbon and Graphite,* London, Society of Chemical Industry, 1982, p. 243.

49. W. N. Reynolds and J. V. Sharp, *Carbon 12,* 103 (1974).

50. S. C. Bennett, D. J. Johnson, and W. Johnson, *J. Mater. Sci. 18,* 3337 (1983).

51. D. J. Johnson, I. Tomizuka, and O. Watanabe, *Carbon 13,* 321 (1975).

52. D. J. Thorne, *Nature 248,* 754 (1974).

53. J. A. C. DaSilva and D. J. Johnson, *J. Mater. Sci. 19,* 3201 (1984).

54. D. J. Johnson, *Chem. Ind.* (London), September 18, 1982, p. 692.

55. W. R. Jones and J. W. Johnson, *Carbon 9,* 645 (1971).

56. M. G. Dobb, D. J. Johnson, and B. P. Saville, *Polymer 22,* 960 (1981).

57. P. D. Ewins and R. T. Potter, *Phil. Trans. R. Soc. Lond. A294,* 507 (1980).

58. S. C. Bennett, D. J. Johnson, and R. Murray, *Carbon 14,* 117 (1976).

59. S. Chwastiak, J. B. Barr, and R. Didchenko, *Carbon 17,* 49 (1979).

60. C. P. Beetz, *Fibre Sci. Technol. 16,* 45 (1982).

61. C. P. Beetz, *Fibre Sci. Technol. 16,* 81 (1982).

2

The Electronic Structure of Graphite and its Basic Origins

MARIE-FRANCE CHARLIER and ALPHONSE CHARLIER

Faculté des Sciences, Université de Metz
Metz, France

INTRODUCTION

The pattern set 38 years ago by P. R. Wallace has had a dominant influence on almost every study of the band theory of graphite up to the present day. This Chapter was undertaken with the intention of presenting a comprehensive account of all the work connected with the tight binding theory. By and large, the theories and doctrines of J. W. McClure, P. R. Wallace, and M. Inoue appear in slightly greater completeness and perhaps in a less attractive form than in the originals. In fact, only a few pages are devoted to the general equations of the electron band theory of solids and a score more to the study of graphite.

Sections I-V contain a general statement of the equations governing free electrons, electrons in a periodic potential, followed by a study of the structure of graphite. In Secs. VI-X, which are also general in character, one will find a discussion of fundamental band models, that is, the tight binding approximation and the definitions of the Slonczewski-Weiss parameters.

The subject of the magnetic Hamiltonian is taken up in Secs. XI-XIV, which deal with the equivalence formulas of Luttinger and Kohn and the Hamiltonian of graphite in the presence of a constant magnetic field. Finally, in Secs. XV-XVIII, which contain the most important results for Landau levels, we investigate the effect of the magnetic field on the energy levels and the calculus of density of states.

GRAPHITE: THEORETICAL STUDY

I. PERIODIC POTENTIAL

A. Free Electron Model [1,2]

1. Direct and Reciprocal Space Lattice

A crystalline substance possesses a periodic three-dimensional structure. The independent vectors $\vec{a}_1$, $\vec{a}_2$, $\vec{a}_3$ that make it possible, from the position of an atom A, to obtain all the atoms exhibiting the same neighborhood as A (i.e., "equivalent" to A) are the basic vectors of the direct crystal lattice.

The vectors $\vec{a}_\alpha$ (α = 1, 2, 3) are not unitary and, as a rule, are not orthogonal

The volume constructed on $\vec{a}_1$, $\vec{a}_2$, $\vec{a}_3$ is the unit cell of the direct space lattice. Its volume is

$$\Omega = (\vec{a}_1 \wedge \vec{a}_2) \cdot \vec{a}_3 = (\vec{a}_1, \vec{a}_2, \vec{a}_3) \tag{1}$$

In the lattice, a point M is identified from an origin O by the vector

$$\overrightarrow{OM} = \vec{r} = \sum_{\alpha=1}^{3} x_\alpha \vec{a}_\alpha \tag{2}$$

If the x_α are all whole numbers, according to the foregoing, point M defines the position of an atom or lattice point.

The study of wave functions introduces terms of the form $e^{i\vec{k}\cdot\vec{r}}$ into the calculations, where $\vec{k}$ is the wave vector of dimension L^{-1}: the argument of the exponential must have the dimension of a pure number. To simplify the calculation of the scalar products $\vec{k}\cdot\vec{r}$, one proceeds by defining new basic vectors.

By setting

$$\vec{b}_1 = \frac{2\pi}{\Omega} \vec{a}_2 \wedge \vec{a}_3$$

$$\vec{b}_2 = \frac{2\pi}{\Omega} \vec{a}_3 \wedge \vec{a}_1 \tag{3}$$

$$\vec{b}_3 = \frac{2\pi}{\Omega} \vec{a}_1 \wedge \vec{a}_2$$

these vectors have the dimension L^{-1}. They are perpendicular to the coordinate planes of the direct space lattice.

In this base, the vector $\vec{k}$ is expressed by

$$\vec{k} = \sum_{\alpha=1}^{3} k_\alpha \vec{b}_\alpha \tag{4}$$

where the k_α are dimensionless numbers. The expression "reciprocal lattice point" is applied to the points obtained by translations of the vectors $\vec{b}_1$, $\vec{b}_2$, and $\vec{b}_3$, with the k_α as whole numbers.

As a rule, the basic vectors of the reciprocal lattice are neither unitary nor orthogonal.

The volume of the elementary unit cell constructed on $\vec{b}_1$, $\vec{b}_2$, $\vec{b}_3$ is

$$\Omega_r = (\vec{b}_1, \vec{b}_2, \vec{b}_3) = \frac{8\pi^3}{\Omega} \tag{5}$$

2. Born's Cyclic Conditions [3,4]

To describe the states of the electrons of a crystal, it is necessary to resolve Schrödinger's equation:

$$\hat{H} \mid \Psi(\vec{r}, t) > = E \mid \Psi(\vec{r}, t) > \tag{6}$$

where $\Psi(\vec{r}, t)$ is the time-dependent wave function. $\hat{H}$ is Hamilton's operator such that

$$\hat{H} = -\frac{\hbar^2}{2m} \nabla^2 + V \tag{7}$$

E is the total mechanical energy.

For the stationary states,

$$\Psi = \Psi_n(\vec{r}, t) = e^{-iE_n t/\hbar} \phi_n(\vec{r}) \tag{8}$$

with ϕ_n the solution of

$$\hat{H} \mid \phi_n(\vec{r}) > = E_n \mid \phi_n(\vec{r}) > \tag{9}$$

E_n is the eigenvalue of the $\hat{H}$ operator associated with the eigenvector $\phi_n(\vec{r})$.

The potential of the crystal lattice is periodic, so that for any translation vector,

$$\vec{T} = \sum_{\alpha=1}^{3} n_\alpha \vec{a}_\alpha \tag{10}$$

where n_α is a whole number, we have

$$V(\vec{r} + \vec{T}) = V(\vec{r}) \tag{11}$$

Explicitly, Eq. (9) is written

$$\left[-\frac{\hbar^2}{2m}\nabla^2 + V(\vec{r})\right] \mid \phi(\vec{r}) > = E \mid \phi(\vec{r}) > \tag{12}$$

As a first approximation, $V(\vec{r})$ can be replaced by its mean value V_0: this is the so-called free electron model.

In the case of motion along a single axis, the solution to Schrödinger's equation is given by

$$\phi''(x) = -k_x^2 \phi(x) \tag{13}$$

where

$$k_x^2 = \frac{2m}{\hbar^2}(E_x - V_0) \tag{14}$$

If $E_x > V_0$, the solution of this is

$$\phi(x) = Ae^{\pm ik_x x} \tag{15}$$

By generalizing to three dimensions

$$\hat{H} = \hat{H}_x + \hat{H}_y + \hat{H}_z \tag{16}$$

and

$$\phi(\vec{r}) = \phi(x)\phi(y)\phi(z) \tag{17}$$

the solution obtained is

$$\phi_{\vec{k}}(\vec{r}) = \phi_0 e^{i\vec{k}\cdot\vec{r}} \tag{18}$$

However

$$\nabla^2\phi_{\vec{k}}(\vec{r}) = -k^2\phi_{\vec{k}}(\vec{r}) \tag{19}$$

and, by substituting in (12),

$$E_k = \frac{\hbar^2 k^2}{2m} + V_0 \tag{20}$$

Solution (18), the plane wave $e^{i\vec{k}\cdot\vec{r}}$, is the solution to Schrödinger's equation for an infinite medium. In actual fact, the crystal volume is limited and the electron cannot escape. One could add the supplementary condition

$$e^{i\vec{k}\cdot\vec{r}}_s = 0 \tag{21}$$

where $\vec{r}_s$ is the position vector of a lattice point of the crystal surface. The writing of continuity conditions of the wave function and of its first derivative in quantum mechanics leads to the quantification of energy.

Born and Von Karman preferred imaginary and cyclic conditions to Eq. (21). The solid is assumed to be constructed by the translation of giant unit cells with sides $2N_1|\vec{a}_1|$, $2N_2|\vec{a}_2|$, $2N_3|\vec{a}_3|$.

The cyclic conditions [3,4] equivalent to the real problem are obtained by writing that the physical properties of the medium are identical at $\vec{r}$ and $\vec{r}'$ such that

$$\vec{r}' = \vec{r} + 2N_1\vec{a}_1 + 2N_2\vec{a}_2 + 2N_3\vec{a}_3 \tag{22}$$

According to (3),

$$\vec{a}_1 \cdot \vec{b}_j = 2\pi \, \delta_{ij} \tag{23}$$

and with (2) and (4),

$$\vec{k} \cdot \vec{r}' = \vec{k} \cdot \vec{r} + 2\pi[2N_1k_1 + 2N_2k_2 + 2N_3k_3] \tag{24}$$

To satisfy

$$e^{i\vec{k}\cdot\vec{r}} = e^{i\vec{k}\cdot\vec{r}'} \tag{25}$$

it is necessary for

$$k_\alpha = \frac{n_\alpha}{2N_\alpha} \tag{26}$$

where n_α is a whole number. The values of energy E_k in Eq. (20) are no longer continuous but quantified.

B. Bloch's Waves

If we resume Eq. (9):

$$\hat{H} \mid \phi_{\vec{k}}(\vec{r}) > \; = E_k \mid \phi_{\vec{k}}(\vec{r}) >$$

the solution $\phi_{\vec{k}}(\vec{r})$ is an eigenfunction of the Hamilton operator $\hat{H}$, but since points $\vec{r}$ and $\vec{r} + \vec{T}$ are equivalent if

$$\vec{T} = n_1\vec{a}_1 + n_2\vec{a}_2 + n_3\vec{a}_3$$

where the n_α are whole numbers, by definition of the operator $\hat{T}$ we must obtain

$$\hat{T} \mid \phi_{\vec{k}}(\vec{r}) > \; = \; \mid \phi_{\vec{k}}(\vec{r} + \vec{T}) > \tag{27}$$

The expression of Bloch's wave function is applied to a solution of the type

$$\phi_{\vec{k}}(\vec{r}) = u_{\vec{k}}(\vec{r}) e^{i\vec{k}\cdot\vec{r}} \tag{28}$$

with

$$u_{\vec{k}}(\vec{r} + \vec{T}) = u_{\vec{k}}(\vec{r}) \tag{29}$$

$u_{\vec{k}}(\vec{r})$ is thus periodic, and if we evaluate the scalar product

$$< \vec{r} \mid \hat{T} \mid \phi_{\vec{k}}(\vec{r}) > \; = \; < \vec{r} \mid \phi_{\vec{k}}(\vec{r} + \vec{T}) > \; = \phi_{\vec{k}}(\vec{r} + \vec{T}) \tag{30}$$

by definition of the base $\vec{r}$ consisting of δ functions.

In fact,

$$\phi_{\vec{k}}(\vec{r} + \vec{T}) = u_{\vec{k}}(\vec{r} + \vec{T}) \; e^{i\vec{k}\cdot\vec{r}} \; e^{i\vec{k}\cdot\vec{T}}$$

according to (28), and with the help of (29):

$$\phi_{\vec{k}}(\vec{r} + \vec{T}) = e^{i\vec{k}\cdot\vec{T}} \; \phi_k(\vec{r}) = e^{i\vec{k}\cdot\vec{T}} < \vec{r} \mid \phi_{\vec{k}}(\vec{r}) > \tag{31}$$

By comparing (31) to (30),

$$\hat{T} \mid \phi_{\vec{k}}(\vec{r}) > \; = e^{i\vec{k}\cdot\vec{T}} \mid \phi_{\vec{k}}(\vec{r}) > \tag{32}$$

This equation with eigenvalues indicates that the Bloch functions (28) are eigenfunctions of the translation operator $\hat{T}$ for the eigenvalue $e^{i\vec{k}\cdot\vec{T}}$.

Since the crystalline potential is periodic [see (11)], points $\vec{r}$ and $\vec{r} + \vec{T}$ are equivalent:

$$\hat{H}(\vec{r}) = \hat{H}(\vec{r} + \vec{T}) \tag{33}$$

That is,

$$\hat{H}\hat{T} - \hat{T}\hat{H} = [\hat{H}, \hat{T}] = \hat{0} \tag{34}$$

$\hat{H}$ and $\hat{T}$ interchange, and these operators have a series of eigenstates in common. Any eigenfunction of the Hamiltonian can therefore be expressed in the form of a Bloch function.

According to (34),

$$\hat{H}\hat{T} \mid \phi_{\vec{k}}(\vec{r}) > = \hat{T}\hat{H} \mid \phi_{\vec{k}}(\vec{r}) >$$

and with (27) and (9),

$$\hat{H} \mid \phi_{\vec{k}}(\vec{r} + \vec{T}) > = E_k \mid \phi_{\vec{k}}(\vec{r} + \vec{T}) > \tag{35}$$

This result indicates that $\psi_{\vec{k}}(\vec{r})$ and $\psi_{\vec{k}}(\vec{r} + \vec{T})$ are solutions of Eq. (9) for the same energy value E_k.

C. Summation Equations [5-8]

We often need equations using summations on the vectors of the direct (or reciprocal) lattice.

1. *Summation on the Direct Lattice*

Let us evaluate the quantity

$$S = \sum_{n} e^{i\vec{R}_n \cdot (\vec{k} - \vec{k}')} \tag{36}$$

where $\vec{R}_n$ is a vector of the direct lattice

$$\vec{R}_n = \sum_{\alpha=1}^{3} x_\alpha \vec{a}_\alpha \tag{37}$$

where the x_α are whole numbers. According to Sec. I.A.2, we have two N_α possible values for x_α:

$$-N_\alpha \leqslant x_\alpha \leqslant N_\alpha - 1 \qquad\qquad (38)$$

The edges of the crystal correspond to the boundaries

$$x_\alpha = -N_\alpha \qquad \text{and} \qquad x'_\alpha = N_\alpha - 1 \qquad (39)$$

Setting

$$\vec{k} - \vec{k}' = \vec{s} = \sum_{\alpha=1}^{3} s_\alpha \vec{b}_\alpha \qquad\qquad (40)$$

we obtain

$$S = S_1 S_2 S_3$$

with

$$S_1 = \sum_{x_1=-N_1}^{N_1-1} e^{2\pi i x_1 s_1} = \frac{\sin 2\pi s_1 N_1}{e^{i\pi s_1} \sin \pi s_1} \qquad (41)$$

If s_1 is the form (26), that is,

$$s_1 = \frac{n_1}{2N_1} \qquad\qquad (42)$$

with n_1 a whole number,

$$\sin 2\pi s_1 N_1 = \sin \pi n_1 = 0 \qquad\qquad (43)$$

If $n_1 \neq 0$, the denominator of (41) is not nil and $S_1 = 0$; if $n_1 = 0$, hence $s_1 = 0$, and we have

$$S_1 = \lim_{s_1 \to 0} \frac{\sin 2\pi s_1 N_1}{\sin \pi s_1} = 2N_1 \qquad (44)$$

In conclusion,

$$S = 8N_1 N_2 N_3 = N \qquad \text{if} \qquad s_1 = s_2 = s_3 = 0$$

that is, if $\vec{k} = \vec{k}'$ and $S = 0$, if $\vec{k} \neq \vec{k}'$, which can be written

$$\sum_n e^{i\vec{R}_n \cdot (\vec{k} - \vec{k}')} = N \, \delta_{\vec{k},\vec{k}'} \tag{45}$$

This implies that if $\vec{k} - \vec{k}' = \vec{G}$, where $\vec{G}$ is a vector of which the extremity represents a point of the reciprocal lattice,

$$\vec{R}_n \cdot \vec{G} = 2\pi(x_1 G_1 + x_2 G_2 + x_3 G_3) = 2\pi m \tag{46}$$

where m is a whole number and

$$\sum_n e^{i\vec{R}_n \cdot \vec{G}} = \sum_n (e^{2\pi i m}) = N \tag{47}$$

This result is valid if $\vec{G} = \vec{0}$. In all other cases,

$$\sum_n e^{i\vec{R}_n \cdot \vec{G}} = 0 \tag{48}$$

2. Summation on the Reciprocal Lattice

Now let us calculate

$$T = \sum_{\vec{k}} e^{i\vec{k} \cdot (\vec{R}_n - \vec{R}_{n'})} \tag{49}$$

where $\vec{R}_n$ and $\vec{R}_{n'}$ identify two equivalent sites of the direct lattice. Hence,

$$\vec{R}_n - \vec{R}_{n'} = \sum_{\alpha=1}^{3} x_\alpha \vec{a}_\alpha \tag{50}$$

where x_α are whole numbers.

If we write

$$\vec{k} = \sum_{i=1}^{3} k_i \vec{b}_i \tag{51}$$

then this gives

$$T = T_1 T_2 T_3 \tag{52}$$

if

$$T_1 = \sum_{k_1} e^{2\pi i k_1 x_1} = \sum_{n_1 = -N_1}^{N_1 - 1} e^{2\pi i n_1 x_1 / 2N_1} \tag{53}$$

as the summation is made on what we call below the "first Brillouin zone," that is,

$$k_1 = \frac{n_1}{2N_1} \tag{54}$$

with n_1 a whole number such that

$$-N_1 \leqslant n_1 \leqslant N_1 - 1 \tag{55}$$

It can be seen that

$$T_1 = \frac{\sin \pi x_1}{e^{-i\pi x_1 / 2N_1} \sin \pi x_1 / 2N_1} \tag{56}$$

If $x_1 \neq 0$, $\sin \pi x_1 = 0$ because x_1 is a whole number by assumption, and since the denominator of T_1 is not nil, $T_1 = 0$. If $x_1 = 0$,

$$\lim_{x_1 \to 0} T_1 = \lim_{x_1 \to 0} \frac{\sin \pi x_1}{\sin \pi x_1 / 2N_1} = 2N_1 \tag{57}$$

It may be concluded that

$$\sum_{\vec{k}} e^{i\vec{k} \cdot (\vec{R}_n - \vec{R}_{n'})} = N \, \delta_{n,n'} \tag{58}$$

If $\vec{k}$ is a vector of the reciprocal lattice denoted $\vec{G}$

$$\sum_{\vec{G}} e^{i\vec{G} \cdot (\vec{R}_n - \vec{R}_{n'})} = N \tag{59}$$

In the opposite cases,

$$\sum_{\vec{G}} e^{i\vec{G} \cdot (\vec{R}_n - \vec{R}_{n'})} = 0 \tag{60}$$

with the summation carried out on the first Brillouin zone.

D. Structural Factor

In a crystal, let us denote by O the origin of a system of identification of the different cells making it up. Let us assume that
each unit cell is a multiple, of order m, and let us denote by A_1,
A_2, . . . , A_m the m atoms of the cell.

In the cell we can identify the different atoms A_α in relation
to A_1. We can write

$$\vec{A}_\alpha = \overrightarrow{A_1 A_\alpha} \tag{61}$$

If $\overrightarrow{OA_1} = \vec{\ell}$ (Fig. 1) and if M is any point of the cell,

$$\overrightarrow{A_\alpha M} = \overrightarrow{A_\alpha A_1} + \overrightarrow{A_1 O} + \overrightarrow{OM} = -\vec{A}_\alpha - \vec{\ell} + \vec{r} \tag{62}$$

The potential is a periodic function in the direct space [see Eq.
(11)], and is hence developable in a Fourier series.

Any wave function Ψ may be written in "$\vec{p}$ base," the base of the
eigenfunctions of the pulse operator $\hat{p}$ or the base of the plane wave
functions. A development in plane wave series can be written

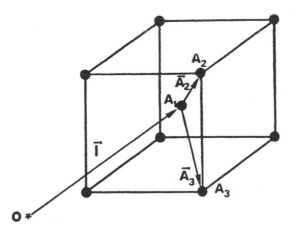

FIG. 1 Identification of lattice points and units cells. O = origin
of crystal; A_1 = origin for unit cell.

$$\Psi = \sum_{\vec{k}} a(\vec{k}) e^{-i\vec{k}\cdot\vec{r}} \tag{63}$$

with

$$a(\vec{k}) = \frac{1}{2\pi} \int_{\Omega} \Psi\, e^{i\vec{k}\cdot\vec{r}}\, d^3r \tag{64}$$

In (63) the summation is carried out on the first Brillouin zone, and in (64) the integration is carried out on the elementary unit cell of the direct lattice whose volume is Ω.

Now, returning to the study of the potential at point M, at this point the potential is the sum of potentials due to the presence of m atoms and ℓ unit cells, or

$$V(\overrightarrow{A_{\alpha}M}) = \sum_{\vec{\ell},\alpha} V_a(\vec{r} - \vec{\ell} - \vec{A}_{\alpha}) \tag{65}$$

where V_a is an atomic potential. The Fourier series development coefficients in reciprocal space are given by (63) and (64), so that

$$V(\overrightarrow{A_{\alpha}M}) = \sum_{\vec{G}} V_{\vec{G}}\, e^{-i\vec{G}\cdot\vec{r}} \tag{66}$$

and

$$V_{\vec{G}} = \frac{1}{\Omega} \int_{\Omega} V(\overrightarrow{A_{\alpha}M}) e^{i\vec{G}\cdot\vec{r}}\, d^3r \tag{67}$$

With the help of (62),

$$V_{\vec{G}} = \frac{1}{\Omega} \sum_{\vec{\ell},\alpha} \int_{\Omega} V_a(\vec{r} - \vec{\ell} - \vec{A}_{\alpha})\, e^{i\vec{G}\cdot\vec{r}}\, d^3r \tag{68}$$

and by changing the origin,

$$\vec{r} \longrightarrow \vec{r}' + \vec{\ell} + \vec{A}_{\alpha} \tag{69}$$

$$V_{\vec{G}} = \frac{1}{\Omega} \sum_{\vec{\ell}} e^{i\vec{G}\cdot\vec{\ell}} \sum_{\alpha} \int_{\Omega} e^{i\vec{G}\cdot\vec{A}_{\alpha}} V_a(\vec{r}')\, e^{i\vec{G}\cdot\vec{r}'}\, d^3r' \tag{70}$$

According to Eq. (47),

$$\sum_{\ell} e^{i\vec{G}\cdot\vec{\ell}} = N$$

and

$$V_{\vec{G}} = \frac{N}{\Omega} \, S_{\vec{G}} \, \int_{\Omega} \, V_a(\vec{r}') \; e^{i\vec{G}\cdot\vec{r}'} \; d^3r' \tag{71}$$

where $S_{\vec{G}}$ is the structural factor

$$S_{\vec{G}} = \sum_{\alpha} e^{i\vec{G}\cdot\vec{A}_{\alpha}} \tag{72}$$

Equation (71) replaces (67). The coefficients $V_{\vec{G}}$ of the Fourier development series of $V(\overrightarrow{A_{\alpha}M})$ are different from zero if $S_{\vec{G}} \neq 0$. According to (47), this condition is satisfied if the vector $\vec{G}$ is a vector of which the extremity is a point of the reciprocal lattice.

By plotting the midperpendicular planes of the vectors that, in reciprocal space, join the origin point to the neighboring points of which the structural factor is not nil, one can construct a polyhedron called the first Jones zone of the crystal.

II. STRUCTURE OF GRAPHITE

Graphite can exhibit a rhombohedral structure, but this is unstable and always poorly crystallized. Our study is limited to hexagonal graphite, of which the structure was determined by Bernal [9] and Maughin [10].

Macroscopically, graphite displays a layered appearance and cleaves fairly easily. This is due to the presence of planes with high atomic density that are relatively distant from each other (3.35 Å) in comparison with the distance between two neighboring carbon atoms belonging to the same plane (d = 1.42 Å).

Each graphite plane displays hexagonal symmetry, and the plane unit cell is defined by two atoms A_0 and B_0 and two crystallographic vectors $\vec{a}_1$ and $\vec{a}_2$ forming an angle of 120° (Fig. 2).

These elementary translations have the modulus

$$a = d \sqrt{3} = 2.46 \text{ Å} \tag{73}$$

If we take A_0 for the origin, it may be noted that the vector $\overrightarrow{A_0B_0}$ cannot be a translation vector of the lattice because it would yield nonexistent lattice points, such as B_0'.

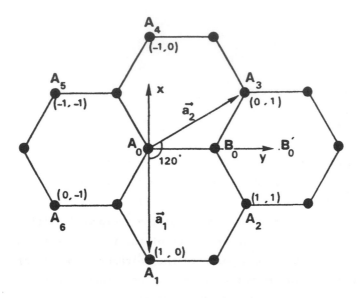

FIG. 2 Plane graphite.

On the other hand, vectors $\overrightarrow{A_0A_1}$ and $\overrightarrow{A_0A_3}$ serve to obtain all the atoms equivalent to A_0, so that half the plane lattice can be constructed. The other half can be obtained by applying the translations $\vec{a}_1$ and $\vec{a}_2$ to atom B_0.

Although all the planes display the same arrangement, the periodicity along the perpendicular c axis is determined by the vector $\vec{a}_3$ with modulus

$$| \vec{a}_3 | = c = 6.708 \text{ Å} \tag{74}$$

The plane thus formed is a second plane next to the first (Fig. 3). In fact, the intermediate plane is offset in relation to its two neighbors by one carbon-carbon bond.

The three-dimensional unit cell (Fig. 4) thus contains four atoms of which the crystallographic nature is different: A_0 and C_0, with a spacing of 3.35 Å on axis c, and B_0 and D_0, which do not have equally close neighbors in this direction.

In the base $\vec{a}_1$, $\vec{a}_2$, $\vec{a}_3$ and according to Fig. 4, the atoms have the following coordinates:

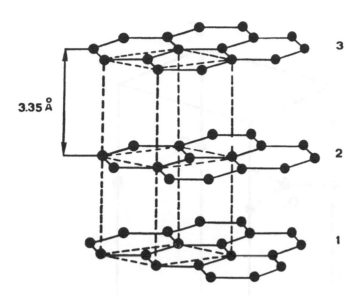

FIG. 3 Graphitic layers.

$$A_0: \quad 0, \ 0, \ 1/4$$
$$C_0: \quad 0, \ 0, \ 3/4$$

(75)

For B_0 one can write

$$B_0: \quad x_B, \ y_B, \ 1/4$$

hence,

$$\overrightarrow{A_0B_0} = x_B\vec{a}_1 + y_B\vec{a}_2$$

and the evaluations of the scalar products

$$\overrightarrow{A_0B_0} \cdot \vec{a}_1 \qquad \text{and} \qquad \overrightarrow{A_0B_0} \cdot \vec{a}_2$$

gave

$$B_0: \quad 1/3, \ 2/3, \ 1/4$$

(76)

Similarly for

$$D_0: \quad 2/3, \ 1/3, \ 3/4$$

(77)

The volume of the unit cell (right-angled prism) is given by the mixed product

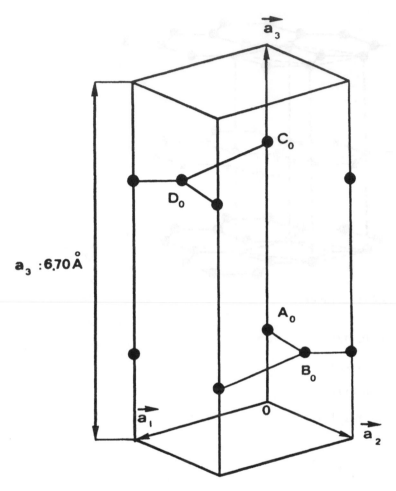

FIG. 4   Three-dimensional unit cell.

$$\Omega = (\vec{a}_1, \ \vec{a}_2, \ \vec{a}_3) \ = \frac{1}{2} \ a^2 c \ \sqrt{3} \qquad\qquad (78)$$

Translation gives the families of atoms A, B, C, and D, which are identified by vectors: $\vec{R}_{A_i}$, $\vec{R}_{B_i}$, and so on. The vector $\vec{r} = x_1 \vec{a}_1 + x_2 \vec{a}_2 + x_3 \vec{a}_3$ denotes any point of the lattice.

III. GROUP OF SYMMETRY OF PLANE GRAPHITE [11-16]

The carbon atom A_0 is taken as the origin of a Cartesian reference system A_0xyz. The unitary vectors along the three axes are $\vec{i}$, $\vec{j}$, and $\vec{k}$, respectively.

Axis A_0z is perpendicular to the plane in Fig. 2, and passage from A_0 to B_0 is carried out by a translation

$$\vec{t} = \frac{a}{\sqrt{3}} \vec{j}$$

We can inventory the following symmetry operations.

The equality denoted E:

$$E = \begin{pmatrix} 1 & 0 & 0 \\ 0 & 1 & 0 \\ 0 & 0 & 1 \end{pmatrix} \qquad (79)$$

The graphitic plane is a plane of symmetry; it retains x and y but changes z to -z:

$$\sigma_h = \begin{pmatrix} 1 & 0 & 0 \\ 0 & 1 & 0 \\ 0 & 0 & -1 \end{pmatrix} \qquad (80)$$

Rotation of 120° about A_0z:

$$C_3 = \begin{pmatrix} -\frac{1}{2} & -\frac{\sqrt{3}}{2} & 0 \\ \frac{\sqrt{3}}{2} & -\frac{1}{2} & 0 \\ 0 & 0 & 1 \end{pmatrix} \qquad (81)$$

Rotation of 240° about A_0z:

$$C_3^2 = \begin{pmatrix} -\frac{1}{2} & \frac{\sqrt{3}}{2} & 0 \\ -\frac{\sqrt{3}}{2} & -\frac{1}{2} & 0 \\ 0 & 0 & 1 \end{pmatrix} \qquad (82)$$

The line A_0B_0 is an axis of symmetry for a rotation of $180°$, and this axis is denoted c_y:

$$c_y = \begin{pmatrix} -1 & 0 & 0 \\ 0 & 1 & 0 \\ 0 & 0 & -1 \end{pmatrix} \tag{83}$$

Axis C_3 is perpendicular to the plane containing c_y, so we also have the axes c_y' and c_y'', with

$$c_y' = C_3 c_y = \begin{pmatrix} \dfrac{1}{2} & -\dfrac{\sqrt{3}}{2} & 0 \\ -\dfrac{\sqrt{3}}{2} & -\dfrac{1}{2} & 0 \\ 0 & 0 & -1 \end{pmatrix} \tag{84}$$

$$c_y'' = C_3^2 c_y = \begin{pmatrix} \dfrac{1}{2} & \dfrac{\sqrt{3}}{2} & 0 \\ \dfrac{\sqrt{3}}{2} & -\dfrac{1}{2} & 0 \\ 0 & 0 & -1 \end{pmatrix} \tag{85}$$

The existence of σ_h and axes C_n perpendicular to the plane σ_h implies the existence of improper axes $S_n = \sigma_h C_n$:

$$S_3 = \sigma_h C_3 = \begin{pmatrix} -\dfrac{1}{2} & -\dfrac{\sqrt{3}}{2} & 0 \\ \dfrac{\sqrt{3}}{2} & -\dfrac{1}{2} & 0 \\ 0 & 0 & -1 \end{pmatrix} \tag{86}$$

$$S_3^2 = \sigma_h C_3^2 = \begin{pmatrix} -\dfrac{1}{2} & \dfrac{\sqrt{3}}{2} & 0 \\ -\dfrac{\sqrt{3}}{2} & -\dfrac{1}{2} & 0 \\ 0 & 0 & -1 \end{pmatrix} \tag{87}$$

Axes of the second order and σ_h create planes of symmetry:

$$\sigma_h c_y = \sigma_d = \begin{pmatrix} -1 & 0 & 0 \\ 0 & 1 & 0 \\ 0 & 0 & 1 \end{pmatrix} \qquad \text{plane } yA_0z \qquad (88)$$

$$\sigma_h c'_y = \sigma'_d = \begin{pmatrix} \dfrac{1}{2} & -\dfrac{\sqrt{3}}{2} & 0 \\[2ex] -\dfrac{\sqrt{3}}{2} & -\dfrac{1}{2} & 0 \\[2ex] 0 & 0 & 1 \end{pmatrix} \qquad (89)$$

$$\sigma_h c''_y = \sigma''_d = \begin{pmatrix} \dfrac{1}{2} & \dfrac{\sqrt{3}}{2} & 0 \\[2ex] \dfrac{\sqrt{3}}{2} & -\dfrac{1}{2} & 0 \\[2ex] 0 & 0 & 1 \end{pmatrix} \qquad (90)$$

The question now arises of how to infer the point symmetry operations for the A atoms from all the operations for the B atoms. By applying a translation,

$$\vec{t} = \vec{j} \frac{a}{\sqrt{3}} = \vec{j} \, d \qquad (91)$$

such that $|\vec{t}| = d = |\overrightarrow{A_0B_0}| = 1.42 \text{ Å}$, carbon sites are created that are nonexistent in reality, whereas if this operation is preceded by an inversion I with center A_0, the type B and type A sites are superposed (Fig. 5).

Since the group of symmetry in this case is the point group, not the space group, it is necessary to add the following symmetry operations:

Inversion I:

$$I = \begin{pmatrix} -1 & 0 & 0 \\ 0 & -1 & 0 \\ 0 & 0 & -1 \end{pmatrix} \qquad (92)$$

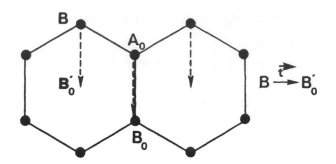

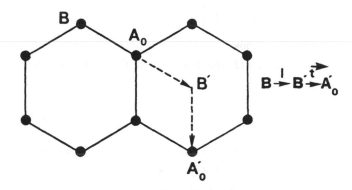

FIG. 5   Passage from type B atoms to type A atoms.

$$IC_3 = \sigma_h C_2 C_3 = \sigma_h C_6^5 = S_6^5:$$

$$S_6^5 = \begin{pmatrix} \dfrac{1}{2} & \dfrac{\sqrt{3}}{2} & 0 \\[2mm] -\dfrac{\sqrt{3}}{2} & \dfrac{1}{2} & 0 \\[2mm] 0 & 0 & -1 \end{pmatrix} \tag{93}$$

$$IC_3^2 = \sigma_h C_2 C_3^2 = \sigma_h C_6^3 C_6^4 = \sigma_h C_6 = S_6:$$

$$S_6 = \begin{pmatrix} \frac{1}{2} & -\frac{\sqrt{3}}{2} & 0 \\ \frac{\sqrt{3}}{2} & \frac{1}{2} & 0 \\ 0 & 0 & -1 \end{pmatrix} \tag{94}$$

We also have

$$Ic_y = \sigma_v = \begin{pmatrix} 1 & 0 & 0 \\ 0 & -1 & 0 \\ 0 & 0 & 1 \end{pmatrix} \tag{95}$$

This is the plane of symmetry $xA_0 z$.

$$Ic_y' = \sigma_v':$$

$$\sigma_v' = \begin{pmatrix} -\frac{1}{2} & \frac{\sqrt{3}}{2} & 0 \\ \frac{\sqrt{3}}{2} & \frac{1}{2} & 0 \\ 0 & 0 & 1 \end{pmatrix} \tag{96}$$

$$Ic_y'' = \sigma_v'' = \begin{pmatrix} -\frac{1}{2} & -\frac{\sqrt{3}}{2} & 0 \\ -\frac{\sqrt{3}}{2} & \frac{1}{2} & 0 \\ 0 & 0 & 1 \end{pmatrix} \tag{97}$$

The axes of rotation of angle π:

$$I\sigma_d = c_x = \begin{pmatrix} 1 & 0 & 0 \\ 0 & -1 & 0 \\ 0 & 0 & -1 \end{pmatrix} \qquad \text{rotation about } A_0 x \tag{98}$$

$$I\sigma_d' = c_x' = \begin{pmatrix} -\frac{1}{2} & \frac{\sqrt{3}}{2} & 0 \\ \frac{\sqrt{3}}{2} & \frac{1}{2} & 0 \\ 0 & 0 & -1 \end{pmatrix} \tag{99}$$

$$I\sigma_d'' = c_x'' = \begin{pmatrix} -\frac{1}{2} & -\frac{\sqrt{3}}{2} & 0 \\ -\frac{\sqrt{3}}{2} & \frac{1}{2} & 0 \\ 0 & 0 & -1 \end{pmatrix} \tag{100}$$

Axes of the sixth order,

$$IS_3 = I\sigma_h C_3 = C_2 C_3 = C_6^5 = \begin{pmatrix} \frac{1}{2} & \frac{\sqrt{3}}{2} & 0 \\ -\frac{\sqrt{3}}{2} & \frac{1}{2} & 0 \\ 0 & 0 & 1 \end{pmatrix} \tag{101}$$

$$IS_3^2 = C_6 = \begin{pmatrix} \frac{1}{2} & -\frac{\sqrt{3}}{2} & 0 \\ \frac{\sqrt{3}}{2} & \frac{1}{2} & 0 \\ 0 & 0 & 1 \end{pmatrix} \tag{102}$$

An axis of second order:

$$I\sigma_h = C_2 = \begin{pmatrix} -1 & 0 & 0 \\ 0 & -1 & 0 \\ 0 & 0 & 1 \end{pmatrix} \tag{103}$$

It is now possible to compile the multiplication table of the point group of plane graphite (Table 1).

The symmetry classes are obtained by forming the products $a_i \cdot a_j \cdot a_i^{-1}$, where a_i is an element of Table 1 and a_i^{-1} its inverse, such that

$$a_i a_i^{-1} = a_i^{-1} a_i = E \tag{104}$$

(for example, it may be noted that if $a_i = S_3$, $a_i^{-1} = S_3^2$).

E forms a class by itself:

$$(a_i E) a_i^{-1} = (E a_i) a_i^{-1} = E \tag{105}$$

TABLE 1   Multiplication Table of Elements of Symmetry of Plane Graphite

I is a class:

$$(a_i I) a_i^{-1} = (I a_i) a_i^{-1} = I \tag{106}$$

The equality $a_i I = I a_i$ is read on Table 1.

σ_h is a class because

$$\sigma_h a_i = a_i \sigma_h \tag{107}$$

irrespective of i.

Similarly for C_2.

For $a_j = C_3$ we have

$$E C_3 E^{-1} = C_3$$

$$\sigma_h C_3 \sigma_h^{-1} = \sigma_h C_3 \sigma_h = C_3$$

$$C_3 C_3 C_3^{-1} = C_3$$

$$C_3^2 C_3 C_3 = C_3$$

$$c_y C_3 c_y^{-1} = C_3$$

$$c_y' C_3 c_y'^{-1} = C_3^2$$

C_3 and C_3^2 form a class denoted $2C_3$.

By the same procedure, it can be shown that

c_y, c_y', c_y'' constitute class $3C_2'$

S_3, S_3^2 constitute class $2S_3$

σ_d, σ_d', σ_d'' constitute class $3\sigma_d$

S_6, S_6^5 constitute class $2S_6$

σ_v, σ_v', σ_v'' constitute class $3\sigma_v$

c_x, c_x', c_x'' constitute class $3C_2''$

C_6^5 and C_6 constitute class $2C_6$

In conclusion, the plane graphite group denoted D_{6h} possesses twelve classes of symmetry: E, $2C_6$, $2C_3$, C_2, $3C_2'$, $3C_2''$, I, $2S_3$, $2S_6$, σ_h, $3\sigma_d$, and $3\sigma_v$.

The number of irreducible representations is equal to the number of classes. We can now compile the table of characters. The rules to be observed are as follows, if we denote $\Gamma^{(i)}$ as the ith irreducible representation.

m_i = order of representation

$\chi^{(i)}(a_j)$ = the spur of the element of symmetry a_j
     in the i representation

$$g = \sum_\alpha m_\alpha^2 \tag{108}$$

where:

g = number of elements (24 in this case)

m_α = whole number

We also have

$$g - \sum_i |\chi^{(\alpha)}(a_i)|^2 \tag{109}$$

$$g\,\delta_{\alpha\beta} = \sum_i [\chi^{(\alpha)}(a_i)]^*\chi^{(\beta)}(a_i) \tag{110}$$

Equation (108) may be written

$$24 = m_1^2 + m_2^2 + m_3^2 + m_4^2 + m_5^2 + m_6^2 + m_7^2 + m_8^2 + m_9^2 + m_{10}^2 + m_{11}^2 + m_{12}^2$$

and we can set

$$m_1 = m_2 = m_3 = m_4 = m_5 = m_6 = m_7 = m_8 = 1 \tag{111}$$

$$m_9 = m_{10} = m_{11} = m_{12} = 2 \tag{112}$$

We have assumed $\Gamma^{(1)} \cdot \cdot \cdot \Gamma^{(8)}$ of the first order. For these irreducible representations E is of the first order, and the spur of E is denoted trE = 1. For $\Gamma^{(9)} \cdot \cdot \cdot \Gamma^{(12)}$, which are of the second order, trE = 2: hence the first column of the table of characters of irreducible representations.

TABLE 2 Characters of Irreducible Representations

	E	$2C_6$	$2C_3$	C_2	$3C_2'$	$3C_2''$	I	$2S_3$	$2S_6$	σ_h	$3\sigma_d$	$3\sigma_v$
$\Gamma_{A_{1g}}^{(1)}$	1	1	1	1	1	1	1	1	1	1	1	1
$\Gamma_{A_{1u}}^{(2)}$	1	1	1	1	1	1	-1	-1	-1	-1	-1	-1
$\Gamma_{A_{2g}}^{(3)}$	1	1	1	1	-1	-1	1	1	1	1	-1	-1
$\Gamma_{A_{2u}}^{(4)}$	1	1	1	1	-1	-1	-1	-1	-1	-1	1	1
$\Gamma_{B_{1g}}^{(5)}$	1	-1	1	-1	1	-1	1	-1	1	-1	1	-1
$\Gamma_{B_{1u}}^{(6)}$	1	-1	1	-1	1	-1	-1	1	-1	1	-1	1
$\Gamma_{B_{2g}}^{(7)}$	1	-1	1	-1	-1	1	1	-1	1	-1	-1	1
$\Gamma_{B_{2u}}^{(8)}$	1	-1	1	-1	-1	1	-1	1	-1	1	1	-1
$\Gamma_{E_{1g}}^{(9)}$	2	1	-1	-2	0	0	2	1	-1	-2	0	0
$\Gamma_{E_{1u}}^{(10)}$	2	1	-1	-2	0	0	-2	-1	1	2	0	0
$\Gamma_{E_{2g}}^{(11)}$	2	-1	-1	2	0	0	2	-1	-1	2	0	0
$\Gamma_{E_{2u}}^{(12)}$	2	-1	-1	2	0	0	-2	1	1	-2	0	0

Since the representation $X^{(1)}(a_i) = 1$ always exists, irrespective of i, we can complete the first line (Table 2). For $\Gamma^{(2)}$ (and up to $\Gamma^{(8)}$) it is necessary to check (109):

$$g = \sum_i |X^{(\alpha)}(a_i)|^2$$

Hence, for example for $\Gamma^{(2)}$,

$$24 = 1 + 2x^2(C_6) + 2x^2(C_3) + x^2(C_2) + 3x^2(C_2') + 3x^2(C_2'') + x^2(I)$$
$$+ 2x^2(S_3) + 2x^2(S_6) + x^2(\sigma_h) + 3x^2(\sigma_d) + 3x^2(\sigma_v)$$

This gives $x^{(2)}(a_i) = \pm 1$ irrespective of i. For $\Gamma^{(9)}(E)$, we have $x^{(9)}(E) = 2$; hence,

$$24 = 2^2 + 2x^2(C_6) + 2x^2(C_3) + x^2(C_2) + 3x^2(C_2') + 3x^2(C_2'') + x^2(I)$$
$$+ 2x^2(S_3) + 2x^2(S_6) + x^2(\sigma_h) + 3x^2(\sigma_d) + 3x^2(\sigma_v)$$

Thus,

$$x^{(9)}(C_6) = \pm 1 \qquad x^{(9)}(C_2') = 0 \qquad x^{(9)}(C_2) = \pm 2$$

$$x^{(9)}(C_3) = \pm 1 \qquad x^{(9)}(C_2'') = 0 \qquad x^{(9)}(I) = \pm 2$$

$$x^{(9)}(S_3) = \pm 1 \qquad x^{(9)}(\sigma_d) = 0 \qquad x^{(9)}(\sigma_h) = \pm 2$$

$$x^{(9)}(S_6) = \pm 1 \qquad x^{(9)}(\sigma_v) = 0$$

and by writing the orthogonality of $\Gamma^{(1)}$ and $\Gamma^{(9)}$ by (110),

$$g\,\delta_{19} = \sum_i [x^{(1)}(a_i)]^* x^{(9)}(a_i)$$

This gives the ninth line of the table.

The representation $\Gamma^{(2)}$ must be orthogonal to $\Gamma^{(1)}$ and $\Gamma^{(9)}$, and so on; hence the filling of the table. The notations used are the following. For example, for $\Gamma^{(1)} \equiv A_{1g}$, A because it is of the first order, g as *gerade* (u as *ungerade*).

A representation is of type g if

$$X(a_i) = X(Ia_i) \tag{113}$$

and of type u if

$$X(a_i) = -X(Ia_i) \tag{114}$$

Hence for $\Gamma^{(1)}$ if $a_i = S_3$, since $IS_3 = C_6^5$ and since $x^{(1)}(S_3) = 1$ and $x^{(1)}(C_6) = 1$, we have a representation g, whereas for $\Gamma^{(2)}$, we have $x^{(2)}(S_3) = -1$, $x^{(2)}(C_6) = 1$, and the representation is of type u.

IV.   SLATER'S ORBITALS [15-21]

In the graphitic plane, each carbon atom is bound to its three neighbors by equivalent σ bonds (Fig. 6), which are localized corresponding to an SP^2 hybridization of the atomic orbitals.

The carbon atom has six electrons distributed over the K and L shells. Two electrons 1S ($\ell = 0$) are found on the K shell (n = 1), and four electrons are found on the L shell (n = 2), of which two are in state 2S ($\ell = 0$) and two in state 2P ($\ell = 1$, m = 0, ±1).

n is the principal quantum number:

$$n = 1, \ 2, \ 3, \ 4, \ . \ . \ . \tag{115}$$

ℓ is the azimuthal quantum number:

$$\ell = 0, \ 1, \ 2, \ . \ . \ . \ , \ n - 1 \tag{116}$$

m is the magnetic quantum number:

$$m = 0, \ \pm 1, \ \pm 2, \ . \ . \ . \ , \ \pm \ell \tag{117}$$

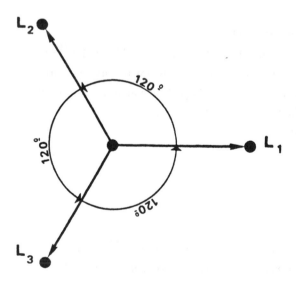

FIG. 6   Trigonal hybridization of carbon.

The wave functions or atomic orbitals [17,18] corresponding to these
states are obtained by multiplying the spherical harmonics $Y_\ell^m(\theta, \phi)$
by the radial functions $R_{n,\ell}(r)$ determined from those of the hydrogen atom, taking account of a charge Z - s, where Z is the number of
protons (six for carbon) and s is a screening constant that depends
on the electrons considered.

Let us consider the electron j of an atom. Slater states that
everything occurs as if the other electrons, noted i with i ≠ j,
situated on

The same electron shell as j

Lower shells, hence closer to the nucleus

play the role of a screen in relation to the nucleus and reduce the
attraction exerted by the latter on the j electron (the electrons
located on shells above j do not form a screen and are not involved
in the calculation).

Slater replaces the charge Ze of the nucleus by a value $Z_j'e$,
called the "apparent" charge of the nucleus in relation to the j
electron considered and poses

$$Z_j' = Z - \sum_i \sigma_{ij} \qquad i \neq j \tag{118}$$

where σ_{ij} is the screen coefficient of one of the i electrons. A
complete theory helps to compile Table 3.

To improve further the agreement between the numerical values
of the real energy levels and the theoretical values, Slater replaces
the principal quantum number n by the apparent quantum number n',
such that

n	1	2	3	4	5	6
n'	1	2	3	3.7	4	4.2

for the carbon atom Z = 6 and the electron configuration is
$1s^2 2s^2 2p^2$.

TABLE 3   Screen Coefficients σ_{ij}

State of j electron	State of i electron						
	1s	2s 2p	3s 3p	3d	4s 4p	4d	4f
1s	0.31						
2s 2p	0.85	0.35					
3s 3p	1	0.85	0.35				
3d	1	1	1	0.35			
4s 4p	1	1	0.85	0.85	0.35		
4d	1	1	1	1	1	0.35	
4f	1	1	1	1	1	1	0.35

Case of two electrons 1s ($n = n' = 1$): the only screen is due to the other electron 1s; hence,

$$\sigma_{1s} = \sum_i \sigma_{ij} = 0.31 \qquad i \neq j$$

and

$$Z'_{1s} = Z - \sigma_{1s} = 5.69$$

Case of electrons 2s and 2p: the screen coefficient is the same for these four electrons ($n = n' = 2$). We must take account of the presence of the two 1s electrons; hence, $\sigma_1 = 2 \times 0.85 = 1.7$, and of the presence, for each of these four electrons, of three other electrons 2s and 2p: $\sigma_2 = 3 \times 0.35 = 1.05$, and hence,

$$Z'_{2s,2p} = 6 - (\sigma_1 + \sigma_2) = 3.25 \tag{119}$$

The radial functions are given by

$$R_{10}(r) = 2\alpha^{3/2}e^{-\alpha r} \qquad (120)$$

$$R_{20}(r) = 2\beta^{3/2}(1 - \beta r)e^{-\beta r} \qquad (121)$$

$$R_{21}(r) = \frac{2}{\sqrt{3}}\beta^{5/2}re^{-\beta r} \qquad (122)$$

where

$$\alpha = \frac{Z - \sigma_{1s}}{a_0} = 10.76 \text{ Å}^{-1} \qquad (123)$$

$$\beta = \frac{Z - (\sigma_1 + \sigma_2)}{2a_0} = 3.07 \text{ Å}^{-1} \qquad (124)$$

and a_0 is Bohr's radius:

$$a_0 = 0.529 \text{ Å} \qquad (125)$$

For the spherical harmonics,

$$Y_0^0 = \frac{1}{2\sqrt{\pi}} \qquad (126)$$

$$Y_1^{-1} = \sqrt{\frac{3}{8\pi}} \sin\theta \; e^{-1\psi} \qquad (127)$$

$$Y_1^0 = \sqrt{\frac{3}{4\pi}} \cos\theta \qquad (128)$$

$$Y_1^1 = \sqrt{\frac{3}{8\pi}} \sin\theta \; e^{i\phi} \qquad (129)$$

This gives the wave functions for the different states.

For the S states,

$$X_{1s} = \frac{\alpha^{3/2}}{\sqrt{\pi}} \; e^{-\alpha r} \qquad (130)$$

$$X_{2s} = \frac{\beta^{3/2}}{\sqrt{\pi}} (1 - \beta r)e^{-\beta r} \qquad (131)$$

For the three 2P states,

$$X_{2P}^{-1} = \frac{\beta^{5/2}}{\sqrt{2\pi}} \, re^{-\beta r} \, \sin\theta \, e^{-i\phi} \tag{132}$$

$$X_{2P}^{0} = \frac{\beta^{5/2}}{\sqrt{\pi}} \, re^{-\beta r} \, \cos\theta = X_{2P_z} \tag{133}$$

$$X_{2P}^{1} = -\frac{\beta^{5/2}}{\sqrt{2\pi}} \, re^{-\beta r} \, \sin\theta \, e^{i\phi} \tag{134}$$

It may be noted that the function X_{2P}^{0} (133) is proportional to $r\cos\theta$, that is, to the expression of z in spherical coordinates, hence the notation X_{2P_z} (Fig. 7).

Similarly, linear combinations of X_{2P}^{-1} and X_{2P}^{1} serve to identify x and y and to define

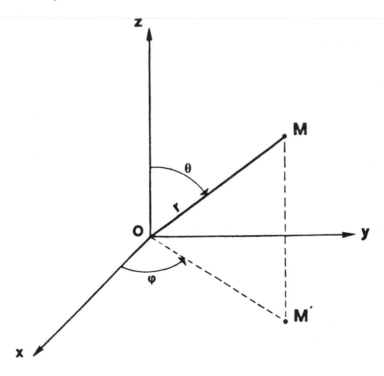

FIG. 7 Spherical coordinates: $x = r \sin\theta \cos\phi$; $y = r \sin\theta \sin\phi$; $z = r \cos\theta$.

$$X_{2P_x} = \frac{1}{\sqrt{2}} \; (X_{2P}^{-1} - X_{2P}^{1}) = \frac{\beta^{5/2}}{\sqrt{\pi}} \; e^{-\beta r} \; r \; \sin \theta \; \cos \phi \tag{135}$$

$$X_{2P_y} = \frac{i}{\sqrt{2}} \; (X_{2P}^{-1} + X_{2P}^{1}) = \frac{\beta^{5/2}}{\sqrt{\pi}} \; e^{-\beta r} \; r \; \sin \theta \; \sin \phi \tag{136}$$

As the carbon atoms move closer together, the 2S and 2P levels are split into bands, allowing a redistribution of the electrons in the states (Fig. 8).

The three electrons participating in the 120° bonds in the graphite plane, which are localized σ bonds, correspond to a hybrid state SP^2.

The wave functions associated with these states may be written

$$X_1 = c_1 X_{2S} + c_1' X_{2P_x} + c_1'' X_{2P_y}$$

$$X_2 = c_2 X_{2S} + c_2' X_{2P_x} + c_2'' X_{2P_y} \tag{137}$$

$$X_3 = c_3 X_{2S} + c_3' X_{2P_x} + c_3'' X_{2P_y}$$

The contribution of X_{2S} (spherical orbital) is identical for the three σ bonds:

$$c_1 = c_2 = c_3 = c \tag{138}$$

The functions defined by (137) must be unitary and orthogonal:

$$< X_i \mid X_j > = \delta_{ij} \tag{139}$$

These conditions provide six equations of which the numerical notations are

$$c = \frac{i}{\sqrt{3}} \tag{140}$$

$$c_1' = \sqrt{\frac{2}{3}} \tag{141}$$

$$c_2' = -\frac{1}{\sqrt{6}} = c_3' \tag{142}$$

$$c_2'' = \frac{1}{\sqrt{2}} = -c_3'' \tag{143}$$

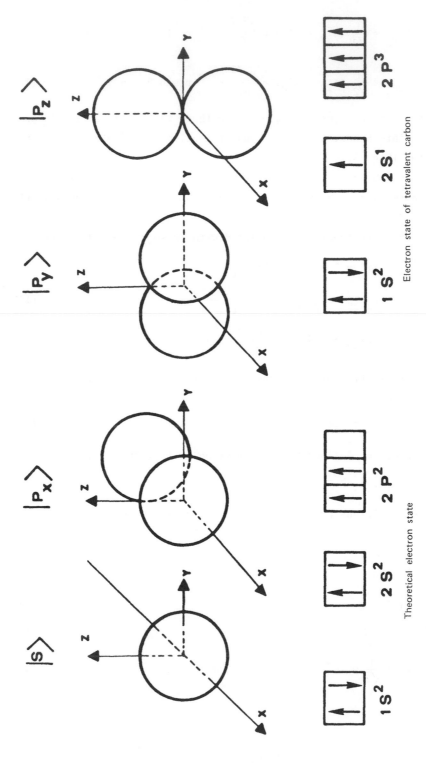

FIG. 8   Electron structure of carbon.

The eigenfunctions (137) are expressed in the form

$$X_1 = \frac{1}{\sqrt{3}} X_{2S} + \sqrt{\frac{2}{3}} X_{2P_x} \tag{144}$$

$$X_2 = \frac{1}{\sqrt{3}} X_{2S} - \frac{1}{\sqrt{6}} X_{2P_x} + \frac{1}{\sqrt{2}} X_{2P_y} \tag{145}$$

$$X_3 = \frac{1}{\sqrt{2}} X_{2S} - \frac{1}{\sqrt{6}} X_{2P_x} - \frac{1}{\sqrt{2}} X_{2P_y} \tag{146}$$

To simplify the resolution of (139), the direction of X_1 was selected to merge with the axis O_x, such that $c_1'' = 0$.

The fourth electron is in a state noted

$$X_4 = X_{2P_z} \qquad \text{see (133)}$$

X_4 is an antisymmetrical function in relation to the graphitic plane and leads to a maximum electron density in the direction perpendicular to this plane.

The fourth electron is a π electron, that is, delocalized.

Now let us try to clarify the form of $u_{\vec{k}}(\vec{r})$ in Eq. (28) if the Bloch wave functions are constructed with Slater's orbitals. If

$$\vec{r} \longrightarrow \vec{r} - \vec{R}_j$$

Eq. (28) is written

$$\phi_{\vec{k}}(\vec{r} - \vec{R}_j) = e^{i\vec{k}\cdot\vec{r}} e^{-i\vec{k}\cdot\vec{R}_j} u_{\vec{k}}(\vec{r} - \vec{R}_j) \tag{147}$$

since $u_{\vec{k}}(\vec{r})$ has the period of the crystal lattice,

$$\phi_{\vec{k}}(\vec{r} - \vec{R}_j) = e^{i\vec{k}\cdot\vec{r}} e^{-i\vec{k}\cdot\vec{R}_j} u_{\vec{k}}(\vec{r}) = e^{-i\vec{k}\cdot\vec{R}_j} \phi_{\vec{k}}(\vec{r}) \tag{148}$$

Hence,

$$\phi_{\vec{k}}(\vec{r}) = e^{i\vec{k}\cdot\vec{R}_j} \phi_{\vec{k}}(\vec{r} - \vec{R}_j) \tag{149}$$

A crystalline wave function may be represented by

$$\phi_{\vec{k}}(\vec{r}) = \sum_j e^{i\vec{k}\cdot\vec{R}_j} \phi_{\vec{k}}(\vec{r} - \vec{R}_j) \tag{150}$$

This function, constructed with the help of atomic orbitals $X_n(\vec{r} - \vec{R}_j)$ in which n characterizes the state of the electron, is

$$\phi_{\vec{k}}(\vec{r}) = \sum_j c_j(\vec{k}) X_n(\vec{r} - \vec{R}_j) \tag{151}$$

By comparing (150) and (151),

$$\phi_{\vec{k}}(\vec{r}) = \sum_j e^{i\vec{k}\cdot\vec{R}_j} X_n(\vec{r} - \vec{R}_j) \tag{152}$$

which we express in the form (147); that is,

$$u_{\vec{k}}(\vec{r}) = \sum_j e^{-i\vec{k}\cdot(\vec{r}-\vec{R}_j)} X_n(\vec{r} - \vec{R}_j) \tag{153}$$

To obtain a crystal function normalized to unity, the following condition must be satisfied:

$$< \phi_{\vec{k}}(\vec{r}) \mid \phi_{\vec{k}}(\vec{r}) > = 1 \tag{154}$$

Hence if

$$< X_n(\vec{r} - \vec{R}_j) \mid X_n(\vec{r} - \vec{R}_\ell) > = \delta_{j\ell} \tag{155}$$

(no orbital overlap), and with (47) this gives

$$\psi_{\vec{k}}(\vec{r}) = \frac{1}{\sqrt{N}} \sum_j e^{i\vec{k}\cdot\vec{R}_j} X_n(\vec{r} - \vec{R}_j) \tag{156}$$

V. BRILLOUIN ZONE OF GRAPHITE [22-24]

A. Reciprocal Space Lattice of Graphite

The discussion in Sec. I.A noted the convenience of using vectors of dimension L^{-1}: $\vec{b}_1$, $\vec{b}_2$, $\vec{b}_3$, which serve as a basis to express the wave vectors

$$\vec{k} = k_1\vec{b}_1 + k_2\vec{b}_2 + k_3\vec{b}_3$$

It may be noted that, according to Eq. (3),

$$\vec{a}_i \cdot \vec{b}_j = 2\pi \, \delta_{ij} \tag{157}$$

This equation facilitates the calculation of scalar products of the type $\vec{k} \cdot \vec{r}$.

In order to determine the $\vec{b}_\alpha$ (α = 1, 2, 3) of graphite, we can use the orthonormalized base in Fig. 2. The unitary vectors along the axes are denoted $\vec{e}_x$, $\vec{e}_y$, and $\vec{e}_z$.

The vectors $\vec{e}_z$ and $\vec{a}_3$ are perpendicular to the plane of the figure. We have

$$\vec{b}_1 = \frac{2\pi}{\Omega} \vec{a}_2 \wedge \vec{a}_3 = \frac{2\pi}{a} \vec{e}_x - \frac{2\pi}{a\sqrt{3}} \vec{e}_y \tag{158}$$

or

$$b_1^2 = \vec{b}_1 \cdot \vec{b}_1 = \frac{16\pi^2}{3a^2} \qquad \text{and} \qquad |\vec{b}_1| = \frac{4\pi}{a\sqrt{3}} \tag{159}$$

Similarly,

$$\vec{b}_2 = \frac{2\pi}{a} \vec{e}_x + \frac{2\pi}{a\sqrt{3}} \vec{e}_y \qquad \text{and} \qquad |\vec{b}_2| = |\vec{b}_1| \tag{160}$$

$$\vec{b}_3 = \frac{2\pi}{c} \vec{e}_z \qquad \text{thus} \qquad |\vec{b}_3| = \frac{2\pi}{c} \tag{161}$$

The angle between $\vec{b}_1$ and $\vec{b}_2$ is given by the calculation of the scalar product

$$\vec{b}_1 \cdot \vec{b}_2 = |\vec{b}_1| \, |\vec{b}_2| \cos \theta$$

hence, with the help of (158) and (160), θ = 60°.

The vectors $\vec{b}_1$ and $\vec{b}_2$ are perpendicular to the vectors $\vec{a}_2$ and $\vec{a}_1$, respectively, and thus form an angle of 60°. Their modulus is equal to $4\pi/a\sqrt{3}$.

Starting from any origin Γ, and by using the translations $\vec{b}_1$, $\vec{b}_2$, and $\vec{b}_3$, the points of the reciprocal lattice are obtained (Fig. 9).

A lattice point is identified by a vector

$$\vec{G} = h\vec{b}_1 + k\vec{b}_2 + \ell\vec{b}_3 \tag{162}$$

where h, k and ℓ are whole numbers. This vector is perpendicular to the planes (h, k, ℓ) of the direct lattice; h, k, and ℓ are the Miller subscripts used in crystallography and in diffraction.

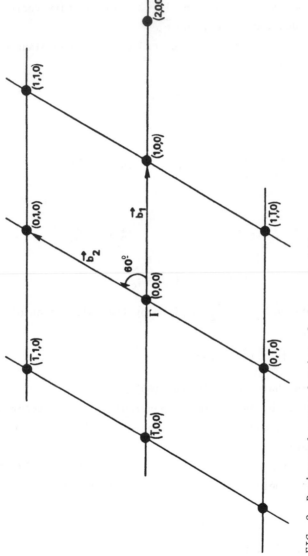

FIG. 9 Reciprocal space lattice of plane graphite.

B.   Jones Zone

Let us evaluate the structural factors

$$S_{\vec{G}} = \sum_{\alpha} e^{i\vec{G}\cdot\vec{A}_{\alpha}} \tag{72}$$

for the points of the reciprocal lattice adjacent to the origin
$\Gamma(0,\ 0,\ 0)$.

> In the plane we find the lattice point $(1,\ 0,\ 0)$ and the five
> equivalent points obtained by a rotation of 60° about an
> axis passing through Γ and perpendicular to the plane of
> Fig. 9: $(1,\ \bar{1},\ 0)$, $(0,\ \bar{1},\ 0)$, $(\bar{1},\ 0,\ 0)$, $(\bar{1},\ 1,\ 0)$ and
> $(0,\ 1,\ 0)$. Point $(1,\ 1,\ 0)$ is excluded: it is a second
> neighbor.
> In the higher plane the equivalent of Γ is the point $(0,\ 0,\ 1)$
> and its equivalents.
> In the lower plane the point $(0,\ 0,\ \bar{1})$ and its equivalents.
> In the *second* higher plane the point $(0,\ 0,\ 2)$ and $(0,\ 0,\ \bar{2})$
> in the *second* lower plane.

Point $(1,\ 0,\ 0)$ corresponds in the upper and lower planes to the
points $(1,\ 0,\ 1)$ and $(1,\ 0,\ \bar{1})$ and their 10 equivalent points:

$$
\begin{aligned}
(1,\ 0,\ 0) &\longrightarrow (1,\ 0,\ 1) \text{ and } (1,\ 0,\ \bar{1}) \\
(1,\ \bar{1},\ 0) &\longrightarrow (1,\ \bar{1},\ 1) \text{ and } (1,\ \bar{1},\ \bar{1}) \\
(0,\ \bar{1},\ 0) &\longrightarrow (0,\ \bar{1},\ 1) \text{ and } (0,\ \bar{1},\ \bar{1}) \\
(\bar{1},\ 0,\ 0) &\longrightarrow (\bar{1},\ 0,\ 1) \text{ and } (\bar{1},\ 0,\ \bar{1}) \\
(\bar{1},\ 1,\ 0) &\longrightarrow (\bar{1},\ 1,\ 1) \text{ and } (\bar{1},\ 1,\ \bar{1}) \\
(0,\ 1,\ 0) &\longrightarrow (0,\ 1,\ 1) \text{ and } (0,\ 1,\ \bar{1})
\end{aligned}
\tag{163}
$$

The vectors $\vec{G}$, which must be considered for the calculation of (72),
are those that correspond to points $(1,\ 0,\ 0)$, $(0,\ 0,\ 1)$, $(1,\ 0,\ 1)$,
and $(0,\ 0,\ 2)$. The vectors $\vec{A}_{\alpha}$ are the four vectors identifying the
atoms

$$
\begin{aligned}
A_0&:\ (0,\ 0,\ 1/4) \\
C_0&:\ (0,\ 0,\ 3/4)
\end{aligned}
\tag{75}
$$

$$B_0: \quad (\tfrac{1}{3}, \tfrac{2}{3}, \tfrac{1}{4}) \tag{76}$$

$$D_0: \quad (\tfrac{2}{3}, \tfrac{1}{3}, \tfrac{3}{4}) \tag{77}$$

Hence the structural factors

$$S_{(1,0,0)} = 1 \tag{164}$$

$$S_{(0,0,1)} = 0 \tag{165}$$

$$S_{(0,0,2)} = -4 \tag{166}$$

$$S_{(1,0,1)} = -\sqrt{3} \tag{167}$$

Considering (165), it is not necessary to take account of points $(0, 0, 1)$ and $(0, 0, \bar{1})$ to construct the Jones zone defined in Sec. I.D.

If we plot the midperpendicular planes of the vectors $\vec{G}$, we obtain Fig. 10 for the plane defined by Γ, $\vec{b}_1$, and $\vec{b}_2$ and Fig. 11 for the plane $(\vec{b}_1, \vec{b}_3)$.

The first Jones zone is a right-angled prism with a hexagonal base of which the edges are slightly truncated. Let us evaluate the volume of the Jones zone. The volume of the pyramid in Fig. 12 is

$$V_P = \tfrac{1}{3} SH \tag{168}$$

The cross section through a plane parallel to the base creates a surface s homothetic to S such that

$$\frac{s}{S} = \left(\frac{h}{H}\right)^2 \tag{169}$$

The volume V_{tr} of the truncated pyramid lying between s and S is given by

$$V_{tr} = \tfrac{1}{3}(SH - sh) \tag{170}$$

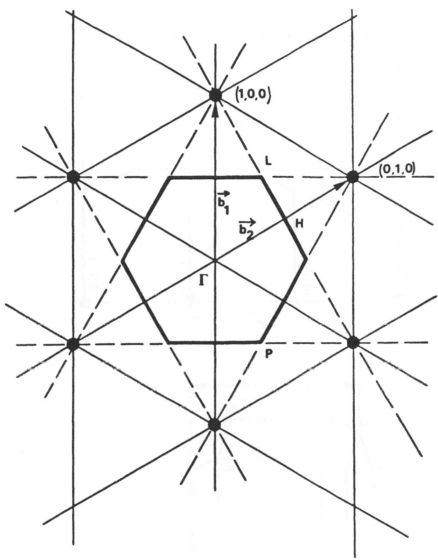

FIG. 10 Jones zone in the plane $(\vec{b}_1, \vec{b}_2)$.

Charlier and Charlier

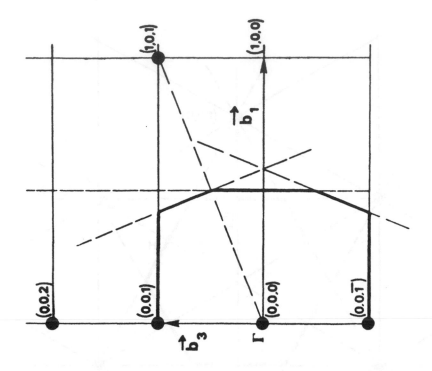

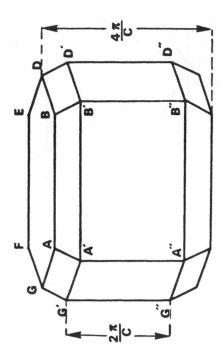

FIG. 11 Jones zone in the plane $(\vec{b}_2, \vec{b}_3)$.

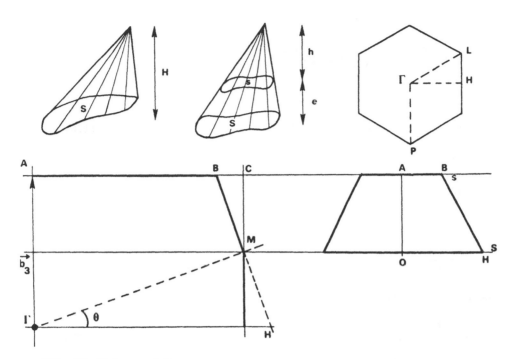

FIG. 12   Volume of Jones zone.

By setting

$$x = \sqrt{\frac{s}{S}} \tag{171}$$

we have

$$x = \frac{h}{H} = \frac{H - e}{H} = 1 - \frac{e}{H} \tag{172}$$

Equation (170) becomes

$$V_{tr} = \frac{eS}{3} \left(\frac{1 - x^3}{1 - x} \right) \simeq \frac{cS}{3} (1 + x + x^2) \tag{173}$$

The area of the hexagon in Fig. 12 is

$$S_{hex} = 3 |\overrightarrow{\Gamma L} \wedge \overrightarrow{\Gamma P}| = 3(\Gamma L)^2 \sin 120° \tag{174}$$

Since

$$\Gamma L = \frac{2\Gamma H}{\sqrt{3}} \qquad S_{hex} = 2(\Gamma H)^2\sqrt{3} \tag{175}$$

However

$$\Gamma H = \frac{b_2}{2} \qquad \text{and } S_{hex} = b_2^2 \frac{\sqrt{3}}{2} \tag{176}$$

Since (Fig. 12)

$$AB = AC - BC = \frac{b_1}{2} - \frac{b_3}{2} \, tg\theta = \frac{b_1}{2}(1 - \alpha^2)$$

If

$$\alpha = \frac{b_3}{b_1}$$

the area s has the value

$$s = \frac{b_1^2\sqrt{3}}{2}(1 - \alpha^2)^2 \tag{177}$$

From (171) we obtain

$$x = 1 - \alpha^2$$

and

$$1 + x + x^2 = 3 - 3\alpha^2 + \alpha^4$$

where $e = b_3/2$. The volume of the truncated pyramid ABDEFGG'A'B'D'E'F'
(Fig. 11) is written

$$V_{tr} = \frac{b_1^2 b_3 \sqrt{3}}{12}(3 - 3\alpha^2 + \alpha^4) \tag{178}$$

The volume V_{pr} of the hexagonal prism A'B'D'E'F'G'G''A''B''E''F'' is

$$V_{pr} = \frac{b_1^2 b_3 \sqrt{3}}{2} \tag{179}$$

and that of the Jones zone,

$$V_{Jones} = V_{pr} + 2V_{tr} = \frac{b_1^2 b_3 \sqrt{3}}{2}\left[2 - \left(\frac{b_3}{b_1}\right)^2 + \frac{1}{3}\left(\frac{b_3}{b_1}\right)^4\right] \tag{180}$$

where

$$b_3 = \frac{2\pi}{c} \qquad \text{and} \qquad b_1 = b_2 = \frac{4\pi}{a\sqrt{3}}$$

C. Brillouin Zone and the Number of States

The following procedure can be used to obtain an elementary unit cell of the direct space lattice. A lattice point is taken as the origin, the lines connecting it to its neighbors are drawn, and the midperpendicular planes of these segments are constructed. The smallest volume lying between the midperpendicular planes is called the Wigner-Seitz cell. The entire space can be filled by means of these cells.

By definition, the Wigner-Seitz cell of the reciprocal lattice is the first Brillouin zone. For graphite, according to Sec. V.B, since point (0, 0, 1) must be considered, this Brillouin zone is a hexagonal prism with height $2\pi/c$; it is the median part of Fig. 11.

We shall now evaluate the number of states, that is, the number of values of $\vec{k}$ in a zone.

The values of the components of the vector $\vec{k}$ are given by (26)

$$k_\alpha = \frac{n_\alpha}{2N_\alpha} \qquad \alpha = 1, 2, 3$$

where n_α is any whole number.

In plane $(\vec{b}_1, \vec{b}_2)$, we obtain Fig. 13. The smallest components are

$$k_1 = \frac{1}{2N_1} \qquad k_2 = \frac{1}{2N_2}$$

They define the vector $\overrightarrow{OK}$. To each extremity of the vector $\vec{k}$ we can associate a parallelogram with the area

$$\sigma = |\overrightarrow{OJ} \wedge \overrightarrow{OI}| = \frac{1}{2N_1} \frac{1}{2N_2} |\vec{b}_1| |\vec{b}_2| \sin 60° = \frac{b_1^2 \sqrt{3}}{8N_1 N_2} \tag{181}$$

With three dimensions, to each extremity we can associate a parallel-epiped of volume

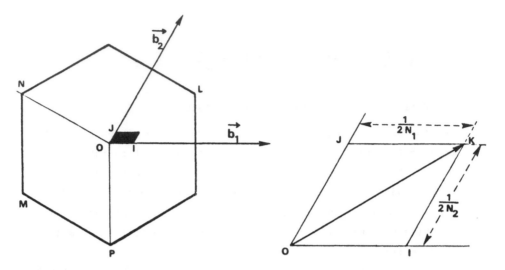

FIG. 13   Number of states.

$$\tau = \frac{b_3}{2N_3} \quad \sigma = \frac{b_1^2 b_3 \sqrt{3}}{2N} \tag{182}$$

if

$$N = 8N_1 N_2 N_3$$

the number of unitary cells in the crystal.

According to (179), the volume of the Brillouin zone is

$$V_B = \frac{b_1^2 b_3 \sqrt{3}}{2} \tag{183}$$

the number of states, that is, of possible extremities of the vector $\vec{k}$, is given by

$$n = \frac{V_B}{\tau} \tag{184}$$

Equations (179) and (182) provide

$$n = N$$

Exactly the same number of permissible values of vectors $\vec{k}$ exist in the first Brillouin zone as in the unit cells in the crystal. A single energy band can contain all the electrons of the graphite.

Taking account of spin and of (183), the number of states in the Jones zone is

$$n = \frac{2V_{Jones}}{\tau} = 2N\left[2 - \left(\frac{b_3}{b_1}\right)^2 + \frac{1}{3}\left(\frac{b_3}{b_1}\right)^4\right]$$

Since

$$b_1 = \frac{4\pi}{a\sqrt{3}} \qquad \text{and} \qquad b_3 = \frac{2\pi}{c}$$

this gives

$$n = 2N\left[2 - \frac{3}{4}\left(\frac{a}{c}\right)^2 + \frac{3}{16}\left(\frac{a}{c}\right)^4\right]$$

Hence for each unit cell there are

$$n' = \frac{n}{N} = 4 - \frac{3}{2}\left(\frac{a}{c}\right)^2 + \frac{3}{8}\left(\frac{a}{c}\right)^4$$

states, or numerically for graphite, 3.80 states per unit cell or 0.95 state per carbon atom.

GRAPHITE: BAND MODELS

VI. TIGHT BINDING APPROXIMATION [5-8]

Based on the atomic orbitals X_{2P_z} given by Eq. (133), centered on the different carbon sites, which we note

$$\phi_i = \phi(\vec{r} - \vec{R}_i) \qquad i = A, B, C, D \qquad (185)$$

we can construct four Bloch functions associated with the families of atoms A, B, C, and D defined in Sec. II:

$$\Psi_j(\vec{r}) = \frac{1}{\sqrt{N}} \sum_i e^{i\vec{k}\cdot\vec{R}_i} \phi_i \qquad j = A, B, C, D \qquad (186)$$

the summation is carried out on the N sites inferred from i by the translations $\vec{a}_1$, $\vec{a}_2$, $\vec{a}_3$. N is the number of unit cells, which is assumed to be very large.

The distance between the energy bands π and σ is not known accurately, but the separation is sufficiently wide to prevent electron transitions between the two bands. Hence one can reasonably ignore the action of the σ electrons on the band structure (see also Coulson [25]).

The desired wave function, which is a crystal wave function of the Bloch wave type and associated with a π electron, is the linear combination of the four functions defined by (186):

$$\Psi(\vec{r}) = \lambda_A \Psi_A + \lambda_B \Psi_B + \lambda_C \Psi_C + \lambda_D \Psi_D \tag{187}$$

It is an eigenfunction of the Hamilton operator $\hat{H}$:

$$\hat{H} \mid \Psi > = E \mid \Psi > \tag{188}$$

E is the energy of a π electron of graphite.

Let us develop (188) with the help of (187):

$$\sum_j \lambda_j \hat{H} \mid \Psi_j > = E \sum_j \lambda_j \mid \Psi_j > \tag{189}$$

where j = A, B, C, D.

By projecting on the states Ψ_i one obtains

$$\sum_j \lambda_j < \Psi_i \mid \hat{H} \mid \Psi_j > = E \sum_j \lambda_j < \Psi_i \mid \Psi_j > \tag{190}$$

Setting

$$H_{ij} = < \Psi_i \mid \hat{H} \mid \Psi_j > \tag{191}$$

a matrix element between states Ψ_i and Ψ_j, and

$$S_{ij} = < \Psi_i \mid \Psi_j > \tag{192}$$

the overlap between states i and j.

Equation (190) represents a system of four equations with four unknowns λ_A, λ_B, λ_C, and λ_D. To avoid the trivial and uninteresting solution $\lambda_j = 0$, it is necessary and sufficient for the determinant of the coefficients to be nil, and this is written symbolically:

$$\det \parallel H_{ij} - E S_{ij} \parallel = 0 \tag{193}$$

Condition (193) is known by the name of secular determinant or equation. Let us clarify (193):

$$\begin{vmatrix} H_{AA} - ES_{AA} & H_{AB} - ES_{AB} & H_{AC} - ES_{AC} & H_{AD} - ES_{AD} \\ H_{BA} - ES_{BA} & H_{BB} - ES_{BB} & H_{BC} - ES_{BC} & H_{BD} - ES_{BD} \\ H_{CA} - ES_{CA} & H_{CB} - ES_{CB} & H_{CC} - ES_{CC} & H_{CD} - ES_{CD} \\ H_{DA} - ES_{DA} & H_{DB} - ES_{DB} & H_{DC} - ES_{DC} & H_{DD} - ES_{DD} \end{vmatrix} = 0 \quad (194)$$

Note that the complete secular determinant [5-8] is a 16 × 16 determinant, which by the separation of elements corresponding to the σ and π electrons is reduced to Eq. (194). The tight binding approximation, applied to graphite, should involve the orbitals of the 2S, $2P_x$, $2P_y$, and $2P_z$ electrons of the four carbon atoms of the unit cell (Fig. 4).

The group theory does not preclude a mixture of these states, but Johnston [8], through his investigation of the Hall effect, substantiated the basis of Coulson's ideas [25-28], which suggested that the σ and π bands were sufficiently distant to justify consideration of the π band independently of the σ band.

VII. SLONCZEWSKI-WEISS PARAMETERS [29]

The Hamiltonian of the π electron in the graphite lattice is

$$H = T + V(\vec{r}) \tag{195}$$

the sum of the kinetic energy T and the potential energy $V(\vec{r})$. If we denote H_0 the Hamilton function of a $2P_z$ electron in the isolated carbon atom and $U(\vec{r} - \vec{R}_i)$ its potential energy, we have

$$H_0 = T + U(\vec{r} - \vec{R}_i) \tag{196}$$

and

$$H_0 \mid \phi_i > = E_0 \mid \phi_i > \tag{197}$$

where E_0 is the energy of the $2P_z$ electron for an isolated carbon atom. The magnitude

$$H' = V(\vec{r}) - U(\vec{r} - \vec{R}_i) \tag{198}$$

is negative (see Fig. 14) and may be considered a perturbation.

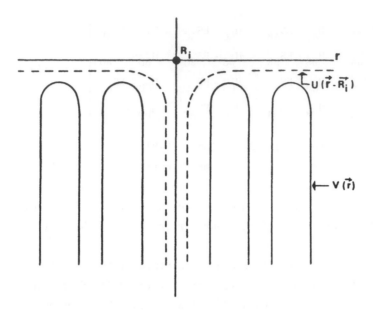

FIG. 14 Potential energies.

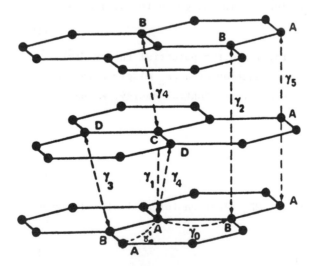

FIG. 15 Slonczewski-Weiss parameters.

The potential energy $V(\vec{r})$ is a periodic function [see Eq. (11)]. Using (198), Slonczewski and Weiss defined the following parameters (Fig. 15).

Interaction between two neighboring atoms A and B:

$$\gamma_0 = -\langle\phi(\vec{r} - \vec{R}_A) \,|\hat{H}'|\, \phi(\vec{r} - \vec{R}_B)\rangle \tag{199}$$

Interaction between two atoms A and C belonging to two adjacent planes:

$$\gamma_1 = \langle\phi(\vec{r} - \vec{R}_A) \,|\hat{H}'|\, \phi(\vec{r} - \vec{R}_C)\rangle \tag{200}$$

Interaction between two B (or D) atoms of second neighboring planes:

$$\gamma_2 = 2\langle\phi(\vec{r} - \vec{R}_B) \,|\hat{H}'|\, \phi(\vec{r} - \vec{R}_B - \vec{a}_3)\rangle \tag{201}$$

Interaction between two B and D atoms of adjacent planes:

$$\gamma_3 = \langle\phi(\vec{r} - \vec{R}_B) \,|\hat{H}'|\, \phi(\vec{r} - \vec{R}_D)\rangle \tag{202}$$

Interaction between two A and D (or B and C) atoms of adjacent planes:

$$\gamma_4 = \langle\phi(\vec{r} - \vec{R}_A) \,|\hat{H}'|\, \phi(\vec{r} - \vec{R}_D)\rangle \tag{203}$$

Interaction between two A (or C) atoms of second neighboring planes:

$$\gamma_5 = 2\langle\phi(\vec{r} - \vec{R}_A) \,|\hat{H}'|\, \phi(\vec{r} - \vec{R}_A - \vec{a}_3)\rangle \tag{204}$$

Also defined are

$$\Delta_A = -\langle\phi(\vec{r} - \vec{R}_A) \,|\hat{H}'|\, \phi(\vec{r} - \vec{R}_A)\rangle \tag{205}$$

$$\Delta_B = -\langle\phi(\vec{r} - \vec{R}_B) \,|\hat{H}'|\, \phi(\vec{r} - \vec{R}_B)\rangle \tag{206}$$

the Coulomb energies of the π electron in the perturbation field H'.

Many authors have suggested a set of parameters determined by experiment. McClure [30] investigated the diamagnetism of graphite and proposed

$$\gamma_0 = 2.8 \text{ eV}$$

$$\gamma_1 = 0.27 \text{ eV} \tag{207}$$

$$\Delta = 0.025 \text{ eV}$$

Based on observations of magnetoreflection and the de Haas-Van Alphen effect, Dresselhaus and Mavroides [31] obtained a value of γ_1 that was much higher and a value of Δ of the opposite sign. Their values are summarized in the first line of the table below.

γ_0	γ_1	γ_2	γ_3	γ_4	γ_5	Δ
2.88	0.395	0.016	0.145	-0.20	0.016	-0.02
2.85	0.31	-0.0185	0.29	0.18	—	—

The second line gives the results of Schroeder et al. [32]. The value of γ_3 is estimated by means of magnetoreflection measurements.

McClure [33] gives the possible intervals of the parameters

$$2.8 < \gamma_0 < 3.2 \text{ eV}$$

$$0.27 < \gamma_1 < 0.40 \text{ eV}$$

$$0.14 < \gamma_3 < 0.29 \text{ eV}$$

$$0.2 < \gamma_4 < 0.3 \text{ eV} \tag{208}$$

$$-0.018 < \gamma_5 = \gamma_2 < -0.02 \text{ eV}$$

$$0.005 < \Delta < 0.1 \text{ eV}$$

We can add the parameter

$$\gamma_0' = <\phi(\vec{r} - \vec{R}_A) \, |\hat{H}'| \, \phi(\vec{r} - \vec{R}_{A'})> \tag{209}$$

where the A' atoms are of the same type as A.

VIII. OVERLAP INTEGRALS

Slater's orbitals of two carbon atoms overlap to varying degrees. Let us assume that the two orbitals $2P_z$ centered at two points O_1 and O_2 are separated by a distance ℓ (Fig. 16). The respective

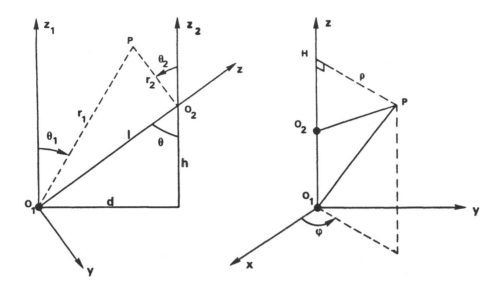

FIG. 16 Spheroidal coordinates.

positions of the two atoms is determined by the angle θ. The current point P is identified in relation to O_i by r_i and θ_i (i = 1, 2).

Let us denote S as the overlap integral

$$S = \frac{\beta^5}{\pi} \int_V r_1 \cos \theta_1 \; r_2 \cos \theta_2 \; e^{-\beta(r_1+r_2)} \; d\tau \qquad (210)$$

Naturally,

$$S = S_{12} = \langle (X_{2P_z})_1 \mid (X_{2P_z})_2 \rangle \qquad (211)$$

where X_{2P_z} is given by (133). This gives, with $h = \ell \cos \theta$,   (212)

$$S = \frac{\beta^5}{\pi} \int_V z_1(z_1 - h) e^{-\beta(r_1+r_2)} \; d\tau \qquad (213)$$

Using the spheroidal coordinates λ, μ, ϕ defined by Fig. 16,

$$\lambda = \frac{r_1 + r_2}{\ell} \qquad 1 \leqslant \lambda < \infty \qquad (214)$$

$$\mu = \frac{r_1 - r_2}{\ell} \qquad -1 \leqslant \mu \leqslant 1 \qquad (215)$$

ϕ angle of rotation about O_{1_z} :

$$0 \leqslant \phi \leqslant 2\pi \tag{216}$$

The volume element $d\tau = dx\ dy\ dz$ becomes

$$d\tau = |J|\ d\mu\ d\lambda\ d\phi \tag{217}$$

where J is the Jacobian

$$J = \begin{vmatrix} \dfrac{\partial x}{\partial \lambda} & \dfrac{\partial x}{\partial \mu} & \dfrac{\partial x}{\partial \phi} \\[2ex] \dfrac{\partial y}{\partial \lambda} & \dfrac{\partial y}{\partial \mu} & \dfrac{\partial y}{\partial \phi} \\[2ex] \dfrac{\partial z}{\partial \lambda} & \dfrac{\partial z}{\partial \mu} & \dfrac{\partial z}{\partial \phi} \end{vmatrix} \tag{218}$$

In the location $(O_1,\ x,\ y,\ z)$ let us express the coordinates of point P. Obviously,

$$x = \quad \cos \phi \tag{219}$$

$$y = \quad \sin \phi \tag{220}$$

and according to (214) and (215),

$$\lambda\mu = \frac{r_1^2 - r_2^2}{\ell^2} \tag{221}$$

Figure 16 makes it possible to write

$$r_1^2 = \rho^2 + z^2 \tag{222}$$

$$r_2^2 = \rho^2 + (z - \ell)^2 \tag{223}$$

hence,

$$r_1^2 - r_2^2 = -\ell^2 + 2\ell z \tag{224}$$

This equation is compared to (221) to obtain

$$z = \frac{\ell}{2}(1 + \lambda\mu) \tag{225}$$

To evaluate the Jacobian, $\rho = PH$ must be expressed as a function of λ, μ, and ϕ. With (214) and (215) let us form the quantity $\lambda^2 + \mu^2$:

$$\lambda^2 + \mu^2 = \frac{2}{\ell^2}(r_1^2 + r_2^2) \tag{226}$$

Taking account successively of (222), (223), (225), this identity becomes

$$\rho^2 = \frac{\ell^2}{4}(\lambda^2 - 1)(1 - \mu^2) \tag{227}$$

Finally, the coordinates of point P are written

$$x = \frac{\ell}{2}\sqrt{(\lambda^2 - 1)(1 - \mu^2)} \cos \phi \tag{228}$$

$$y = \frac{\ell}{2}\sqrt{(\lambda^2 - 1)(1 - \mu^2)} \sin \phi \tag{229}$$

$$z = \frac{\ell}{2}(1 + \lambda\mu) \tag{230}$$

The calculation of (218) is possible, and the result is

$$J = \frac{\ell^3}{8}(\lambda^2 - \mu^2) \tag{231}$$

The purpose of this section is to calculate the integral (213). However,

$$z_\perp = z \cos \theta - y \sin \theta \tag{232}$$

Hence the integration on ϕ amounts to calculating

$$I = \int_0^{2\pi} z_1(z_1 - h) \, d\phi = \pi(2z^2 \cos^2 \theta + \rho^2 \sin^2 \theta - 2hz \cos \theta) \tag{233}$$

By setting $d = \ell \sin \phi$,

$$I = \pi\left[\frac{h^2}{2}(\lambda^2\mu^2 - 1) + \frac{d^2}{4}(\lambda^2 - 1)(1 - \mu^2)\right] \tag{234}$$

and

$$S = \frac{\beta^5 \ell^3}{32} \int_0^1 d\lambda \, e^{-\beta\ell\lambda} \int_{-1}^{+1} (a_4\mu^4 + a_2\mu^2 + a_0) \, d\mu \tag{235}$$

with

$$a_4 = \lambda^2(d^2 - 2h^2) - d^2 \tag{236}$$

$$a_2 = \lambda^4(2h^2 - d^2) + 2h^2 + d^2 \tag{237}$$

$$a_0 = \lambda^4 d^2 - \lambda^2(2h^2 + d^2) \tag{238}$$

After integration on μ,

$$S = \frac{\beta^5 \ell^3}{120} \int_1^\infty e^{-\beta\ell\lambda} [5\ell^2\lambda^4 - 6(\ell^2 + 2h^2)\lambda^2 + (\ell^2 + 4h^2)] \, d\lambda \tag{239}$$

and since

$$I_n = \int_1^\infty \lambda^n e^{-\beta\ell\lambda} \, d\lambda = \frac{n!}{(\beta\ell)^{n+1}} \sum_{j=0}^n \frac{(\beta\ell)^j}{j!} \tag{240}$$

this gives

$$S = e^{-\beta\ell} \left[1 + \beta\ell + \frac{1}{5}\beta^2\ell^2(2 - \cos^2\theta) + \frac{1}{15}\beta^3\ell^3(1 - 3\cos^2\theta) \right.$$

$$\left. - \frac{1}{15}\beta^4\ell^4 \cos^2\theta \right] \tag{241}$$

For the atoms A and B, $\cos\theta = 0$ and $\ell = d$; hence,

$$S_0 = e^{-\beta d} \left(1 + \beta d + \frac{2}{5}\beta^2 d^2 + \frac{1}{15}\beta^3 d^3 \right) \tag{242}$$

For atoms A and C, $\cos\theta = 1$ and $\ell = h = c/2$:

$$S_1 = e^{-\beta\frac{c}{2}} \left(1 + \frac{1}{2}\beta c + \frac{1}{20}\beta^2 c^2 - \frac{1}{60}\beta^3 c^3 - \frac{1}{240}\beta^4 c^4 \right) \tag{243}$$

Let us construct the table of overlap integrals between close neighbors (Table 4).

TABLE 4 Overlap Integrals Between Close Neighbors

h (Å)	$d =$ 0 Å	$d =$ 1.42 Å	$d =$ 2.46 Å	$d =$ 2.84 Å	$d =$ 3.75 Å	$d =$ 4.26 Å
0	1	0.236	3.1×10^{-2}	1.4×10^{-2}	1.6×10^{-3}	4.8×10^{-4}
3.354	-2.9×10^{-2}	-1.4×10^{-2}	-3.5×10^{-3}	-1.8×10^{-3}	-3.2×10^{-4}	-1.1×10^{-4}
6.708	-1.5×10^{-5}	-9.7×10^{-6}	-4.3×10^{-6}	-2.9×10^{-6}	-9.3×10^{-7}	-4.5×10^{-7}

IX.   STRUCTURAL FACTORS OF GRAPHITE

In the secular determinant (194), it is necessary to consider interactions between orbitals and potentials centered on different atoms.
As a rule, *this is limited to the first neighbors.* There are several
equivalent sites for a given atom: for example, an A atom possesses
three B neighbors and six A neighbors in its plane. This gives rise
to structural factors [see Eq. (72)] of the form

$$\sum_{i=1}^{3} e^{i\vec{k}\cdot\overrightarrow{AB}_i}, \qquad \sum_{i=1}^{6} e^{i\vec{k}\cdot\overrightarrow{AA}_i} \tag{244}$$

and so on.

A.   Factor AA

This is identical for the four types of atom. An A (or B, C, or D)
atom possesses six neighbors in the plane (see Fig. 2), such that

$$\overrightarrow{A_0A}_1 = (1,\ 0,\ 0)$$

$$\overrightarrow{A_0A}_2 = (1,\ 1,\ 0)$$

$$\overrightarrow{A_0A}_3 = (0,\ 1,\ 0)$$

$$\overrightarrow{A_0A}_4 = (-1,\ 0,\ 0) \tag{245}$$

$$\overrightarrow{A_0A}_5 = (-1,\ -1,\ 0)$$

$$\overrightarrow{A_0A}_6 = (0,\ -1,\ 0)$$

$$F_1(\vec{k}) = \sum_{i=1}^{6} e^{i\vec{k}\cdot\overrightarrow{A_0A}_i} = e^{2\pi i k_1} + e^{2\pi i (k_1+k_2)} + e^{-2\pi i k_1}$$

$$+ e^{-2\pi i (k_1+k_2)} + e^{-2\pi i k_2} + e^{2\pi i k_2} \tag{246}$$

$$F_1(\vec{k}) = 2 \cos 2\pi\, k_1 + 2 \cos 2\pi\, k_2 + 2 \cos 2\pi\, (k_1 + k_2) \tag{247}$$

For atoms located in the second neighboring planes, we have

$$\overrightarrow{A_0A}_7 = \vec{a}_3 \qquad \overrightarrow{A_0A}_8 = -\vec{a}_3 \tag{248}$$

and

$$\sum_{i=7}^{8} e^{i\vec{k}\cdot\overrightarrow{A_0A_i}} = e^{2\pi i k_3} + e^{-2\pi i k_3} = 2 \cos k_z c \qquad (249)$$

if

$$k_z = \frac{2\pi}{c} k_3 \qquad (250)$$

B. Factors AB and CD

Figure 17 shows that the configurations AB and CD are equivalent. This also applies to the structural factors. We have

$$\overrightarrow{A_0B_1} = \left(\frac{1}{3}, \ \frac{2}{3}, \ 0\right)$$

$$\overrightarrow{A_0B_2} = \left(-\frac{2}{3}, \ -\frac{1}{3}, \ 0\right) \qquad (251)$$

$$\overrightarrow{A_0B_3} = \left(\frac{1}{3}, \ -\frac{1}{3}, \ 0\right)$$

and

$$F_2(\vec{k}) = \sum_{i=1}^{3} e^{i\vec{k}\cdot\overrightarrow{A_0B_i}} = e^{\frac{2\pi i}{3}(k_1+2k_2)} + e^{-\frac{2\pi i}{3}(2k_1+k_2)} + e^{\frac{2\pi i}{3}(k_1-k_2)} \qquad (252)$$

Note that

$$\sum_{i=1}^{3} e^{i\vec{k}\cdot\overrightarrow{C_0D_i}} = \sum_{i=1}^{3} e^{i\vec{k}\cdot\overrightarrow{B_0A_i}} = [F_2(\vec{k})]^* \qquad (253)$$

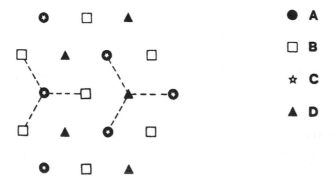

FIG. 17 AB and DC configurations.

C. Factor AC

The two C carbon atoms that are neighbors of A are such that

$$\overrightarrow{A_0C_1} = \frac{\vec{a}_3}{2}$$
$$\overrightarrow{A_0C_2} = -\frac{\vec{a}_3}{2}$$

(254)

and

$$\sum_{i=1}^{3} e^{i\vec{k}\cdot\overrightarrow{A_0C_i}} = e^{i\pi k_3} + e^{-i\pi k_3} = 2\cos\frac{k_z c}{2}$$

(255)

D. Factors AD, CB, and BD

The six D_i carbon atoms that are first neighbors of A_0 are situated on either side of the graphitic plane containing A_0:

$$\overrightarrow{A_0D_i} = \overrightarrow{A_0C_1} + \overrightarrow{C_1D_i} \qquad \text{with } i = 1, 2, 3$$

(256)

for the upper plane, and

$$\overrightarrow{A_0D_i} = \overrightarrow{A_0C_2} + \overrightarrow{C_2D_i} \qquad \text{with } i = 4, 5, 6$$

(257)

for the lower plane.

$$\sum_{i=1}^{3} e^{i\vec{k}\cdot\overrightarrow{A_0D_i}} = e^{i\vec{k}\cdot\overrightarrow{A_0C_1}} \sum_{i=1}^{3} e^{i\vec{k}\cdot\overrightarrow{C_1D_i}} = e^{i\pi k_3} [F_2(\vec{k})]^*$$

(258)

$$\sum_{i=4}^{6} e^{i\vec{k}\cdot\overrightarrow{A_0D_i}} = e^{i\vec{k}\cdot\overrightarrow{A_0C_2}} \sum_{i=4}^{6} e^{i\vec{k}\cdot\overrightarrow{C_2D_i}} = e^{-i\pi k_3} [F_2(\vec{k})]^*$$

(259)

and hence,

$$\sum_{i=1}^{6} e^{i\vec{k}\cdot\overrightarrow{A_0D_i}} = 2\cos\frac{k_z c}{2} [F_2(\vec{k})]^*$$

(260)

Similarly, we obtain

$$\sum_{i=1}^{6} e^{i\vec{k}\cdot\overrightarrow{C_0B_i}} = 2\cos\frac{k_z c}{2} F_2(\vec{k})$$

(261)

$$\sum_{i=1}^{6} e^{i\vec{k}\cdot\overrightarrow{B_0D}_i} = 2 \cos \frac{k_z c}{2} F_2(\vec{k}) \tag{262}$$

Note that $F_2(\vec{k})$ and $F_1(\vec{k})$ can be associated easily simply by using (252) to evaluate the magnitude $F_2^2(k)$:

$$[F_2(\vec{k})]^2 = [F_2(\vec{k})]^* F_2(\vec{k}) = 3 + 2 \cos 2\pi (k_1 + k_2)$$
$$+ 2 \cos 2\pi k_1 + 2 \cos 2\pi k_2 \tag{263}$$

Thus,

$$F_2(\vec{k}) = \sqrt{3 + F_1(\vec{k})} \tag{264}$$

X. BAND MODELS

In chronological order we can consider the work of Wallace [34], Mrozowski [35], and Slonczewski and Weiss [29]. Around 1950, Mrozowski suggested a qualitative model, the culmination of reflections on coke resistivity investigations as a function of the highest treatment temperature (HTT) for pregraphite carbons. In 1953, Mrozowski [35], as the result of resistivity measurements taken on intercalated graphite compounds and by the theoretical interpretation of the properties of graphite at low temperatures, completed his model (Fig. 18). The variation in density of energy levels $n(E)$ of the π band as a function of energy E is given in Fig. 18 for graphite. Most of the models [36] were derived more or less directly from this one.

A. Preliminary Calculations

Let us evaluate the different terms of the secular determinant (194). With the help of (186),

$$\zeta_{ij} = <\Psi_i|\Psi_j> = \frac{1}{N} \sum_{i,j} e^{i\vec{k}\cdot(\vec{R}_j-\vec{R}_i)} <\phi_i|\phi_j> = \frac{1}{N} \sum_{i,j} e^{i\vec{k}\cdot(\vec{R}_j-\vec{R}_i)} S_{ij}$$
$$\tag{265}$$

Note that

$$\zeta_{ij}^* = <\Psi_i|\Psi_j>^* = \zeta_{ji} \tag{266}$$

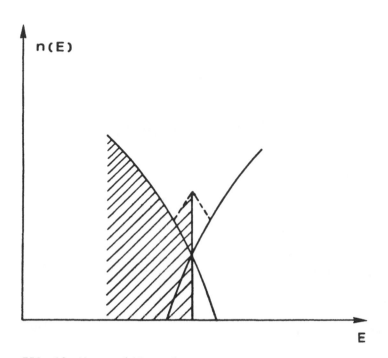

FIG. 18 Mrozowski's model.

An identical equation exists for the matrix elements H_{ij}. In effect the operator $\hat{H}$ is hermetic if

$$\hat{H} = \hat{H}^+ \tag{267}$$

$\hat{H}^+$ is the adjoint operator such that

$$\hat{H}^+ = \tilde{\hat{H}}^* \tag{268}$$

$\tilde{\hat{H}}$ is the transposition of $\hat{H}$, the matrix obtained by interchanging rows and columns; however,

$$H_{ij} = <\Psi_i|\hat{H}|\Psi_j> = \int_V \Psi_i^*\hat{H}\Psi_j \ d\tau \tag{269}$$

and

$$H_{ij} = \int_V \Psi_j (\tilde{\hat{H}}\Psi_i^*) \ d\tau \tag{270}$$

and

$$H_{ij}^* = \int_V \Psi_j^* (\tilde{\hat{H}}^*\Psi_i) \ d\tau = \int_V \Psi_j^*\hat{H}^+\Psi_i \ d\tau = H_{ji} \tag{271}$$

1. Evaluation of ζ_{ij} Terms

According to (265) and (266),

$$\zeta_{AA} = \langle \Psi_A | \Psi_A \rangle = \frac{1}{N} \sum_{A,A'} \langle e^{i\vec{k}\cdot\vec{R}_A} \phi_A | e^{i\vec{k}\cdot\vec{R}_{A'}} \phi_{A'} \rangle$$

$$= 1 + \sum_{A' \neq A} e^{i\vec{k}\cdot(\vec{R}_{A'} - \vec{R}_A)} \langle \phi_A | \phi_{A'} \rangle = 1 + F_1(\vec{k}) S_{AA'} \tag{272}$$

$$\zeta_{AB} = \zeta_{BA}^* = F_2(\vec{k}) S_{AB} \tag{273}$$

$$\zeta_{AC} = \zeta_{CA}^* = 2 S_{AC} \cos \frac{k_z c}{2} \tag{274}$$

$$\zeta_{AD} = \zeta_{DA}^* = 2 S_{AD} \cos \frac{k_z c}{2} \tag{275}$$

By symmetry,

$$\zeta_{AA} = \zeta_{BB} = \zeta_{CC} = \zeta_{DD} \tag{276}$$

We also obtain

$$\zeta_{BC} = \zeta_{CB}^* = 2 F_2(\vec{k}) S_{BC} \cos \frac{k_z c}{2} \tag{277}$$

$$\zeta_{BD} = \zeta_{DB}^* = 2 F_2(\vec{k}) S_{BD} \cos \frac{k_z c}{2} \tag{278}$$

2. Evaluation of H_{ij} Terms

By definition (191), with the help of (186), we have

$$H_{AA} = \langle \Psi_A | \hat{H} | \Psi_A \rangle = \frac{1}{N} \langle \sum_A e^{i\vec{k}\cdot\vec{R}_A} \phi_A | \sum_{A'} e^{i\vec{k}\cdot\vec{R}_{A'}} \phi_{A'} \rangle \tag{279}$$

Hence,

$$N H_{AA} = \sum_{A,A'} e^{i\vec{k}\cdot\overrightarrow{AA'}} \langle \phi_A | \hat{H} | \phi_{A'} \rangle \tag{280}$$

The summation on the A' atoms is the same irrespective of the A atom selected. The summation on A amounts to multiplication by N:

$$H_{AA} = \sum_{A'} e^{i\vec{k}\cdot\overrightarrow{AA'}} \langle \phi_A | \hat{H} | \phi_{A'} \rangle \tag{281}$$

Let us isolate the case A' = A:

$$H_{AA} = <\phi_A|\hat{H}|\phi_A> + \sum_{A'\neq A} e^{i\vec{k}\cdot\vec{AA'}} <\phi_A|\hat{H}|\phi_{A'}> \qquad (282)$$

Since (see Sec. VII)

$$<\phi_A|\hat{H}|\phi_A> = <\phi_A|\hat{H}_0 + \hat{H}'|\phi_A> = E_0 - \Delta_A \qquad (283)$$

and since

$$<\phi_A|\hat{H}|\phi_{A'}> = <\phi_A|\hat{H}_0 + \hat{H}'|\phi_{A'} = E_0 S_{AA'} + <\phi_A|\hat{H}'|\phi_{A'}> \qquad (284)$$

$$H_{AA} = E_0 - \Delta_A + \gamma_5 \cos k_z c + (E_0 S_{AA'} + \gamma_0')F_1(\vec{k}) \qquad (285)$$

if we consider only the two A' atoms determined from $\vec{A}$ by a vector $\vec{a}_3$ (74) for the other planes (250).

Similarly,

$$H_{AB} = <\Psi_A|\hat{H}|\Psi_B> = \frac{1}{N} \sum_{A,B} e^{i\vec{k}\cdot\vec{AB}} <\phi_A|\hat{H}_0 + \hat{H}'|\phi_B>$$

$$= -\gamma_0 F_2(\vec{k}) + E_0 F_2(\vec{k})S_{AB} \qquad (286)$$

$$H_{AC} = 2\gamma_1 \cos \frac{k_z c}{2} + 2E_0 S_{AC} \cos \frac{k_z c}{2} \qquad (287)$$

$$H_{AD} = 2\gamma_4 F_2^*(\vec{k}) \cos \frac{k_z c}{2} + 2E_0 S_{AC} F_2^*(\vec{k}) \cos \frac{k_z c}{2} \qquad (288)$$

$$H_{BD} = 2\gamma_3 F_2(\vec{k}) \cos k_z \frac{c}{2} + 2E_0 F_2(\vec{k})S_{BD} \cos k_z \frac{c}{2} \qquad (289)$$

We have seen in Sec. IX that some structural factors are conjugate. This gives rise to the equalities

$$H_{CD} = H_{AB}^* \qquad (290)$$

$$H_{CB} = H_{AD}^* \qquad (291)$$

For reasons of symmetry (see Fig. 3),

$$H_{CC} = H_{AA} \qquad (292)$$

$$H_{DD} = H_{BB} \qquad (293)$$

As it happens, H_{BB} assumes the same form as H_{AA}. It suffices in (285) to replace Δ_A by Δ_B and γ_5 by γ_2:

$$H_{BB} = E_0 - \Delta_B + \gamma_2 \cos k_z c + (E_0 S_{AA'} + \gamma_0') F_1(\vec{k}) \tag{294}$$

B. Plane Model [34,37,39]

We can add the following simplifying hypothesis: the distance between graphitic planes is sufficiently greater than the carbon-carbon distance in the plane to assume that the different graphite layers are independent.

The determinant (194) becomes

$$\begin{vmatrix} H_{AA} - E\zeta_{AA} & H_{AB} - E\zeta_{AB} \\ H_{AB}^* - E\zeta_{AB}^* & H_{BB} - E\zeta_{BB} \end{vmatrix} = 0 \tag{295}$$

with

$$\zeta_{AA} = \zeta_{BB} = 1 \tag{296}$$

$$\zeta_{AB} = F_2(k) S_{AB} \tag{297}$$

$$H_{AA} = E_0 - \Delta_A + (E_0 S_{AA'} + \gamma_0') F_1(k) \tag{298}$$

$$H_{AB} = (E_0 S_{AB} - \gamma_0) F_2(k) \tag{299}$$

The resolution of (295) gives

$$E = \frac{H_{AB}\zeta_{AB}^* + H_{AB}^*\zeta_{AB} - 2H_{AA} \pm \sqrt{(H_{AB}\zeta_{AB}^* + H_{AB}^*\zeta_{AB} - 2H_{AA})^2 - 4(1 - |\zeta_{AB}|)^2 [|H_{AA}|^2 - |H_{AB}|^2]}}{2[1 - |\zeta_{AB}|^2]} \tag{300}$$

By substitution,

$$E_+ = E_0 + \frac{\gamma_0 |F_2(\vec{k})| + (E_0 S_{AA'} + \gamma_0') F_1(\vec{k}) - \Delta_A}{1 - S_{AB}|F_2(\vec{k})|} \tag{301}$$

$$E_- = E_0 - \frac{\gamma_0 |F_2(\vec{k})| - (E_0 S_{AA'} + \gamma_0') F_1(\vec{k}) + \Delta_A}{1 + S_{AB}|F_2(\vec{k})|} \tag{302}$$

If we add Hückel's condition, that is, assuming that the orbitals do not overlap,

$$\langle \phi_i | \phi_j \rangle = S_{ij} = \delta_{ij} \tag{303}$$

we obtain the formulas proposed by Wallace:

$$E_+ = E_0 + \gamma_0 |F_2(\vec{k})| + \gamma_0' F_1(\vec{k}) - \Delta_A \tag{304}$$

$$E_- = E_0 - \gamma_0 |F_2(\vec{k})| - \gamma_0' F_1(\vec{k}) + \Delta_A \tag{305}$$

Figure 19 represents the Brillouin zone of graphite with three dimensions. The notation of the points in the zone is that of Koster [39].

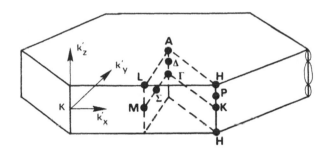

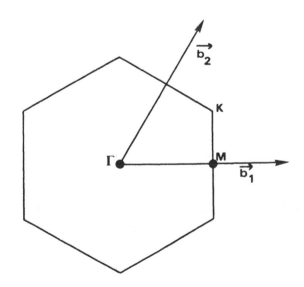

FIG. 19 Koster's notation.

For the plane model, let us evaluate the energy at a few specific points. The energy E_- corresponds to points inside the zone and E_+ to points outside it. Hence, at the center Γ, $k_1 = k_2 = 0$; thus,

$$F_1(\vec{k}) = 2 \cos 2\pi k_1 + 2 \cos 2\pi k_2 + 2 \cos 2\pi (k_1 + k_2) \quad (247)$$

$$|F_2(\vec{k})| = \sqrt{3 + F_1(\vec{k})}$$

thus,

$$F_1(\vec{k}) = 6 \tag{306}$$

and

$$E_\Gamma = E_0 - 3\gamma_0 - 6\gamma_0' + \Delta_A \tag{307}$$

At M, $k_1 = 1/2$, and $k_2 = 0$:

$$F_1(\vec{k}) = -2 \tag{308}$$

and

$$E_M = E_0 - \gamma_0 + 2\gamma_0' + \Delta_A \tag{309}$$

At the corner K of the zone, $k_1 = k_2 = 1/3$:

$$F_1(\vec{k}) = -3 \tag{310}$$

and

$$E_K = E_0 + 3\gamma_0' + \Delta_A \tag{311}$$

In the Brillouin zone, we can now plot the isoenergy curves (Fig. 20). The energy E_- given by (304) depends only on the length of the vector $\vec{k}$. To find the curves

$$E_- = \text{constant} \tag{312}$$

it is necessary to find, from (304) and (264), the locus of the points, extremities of the vector $\vec{k}$, such that

$$F_1(\vec{k}) = \text{constant} \tag{313}$$

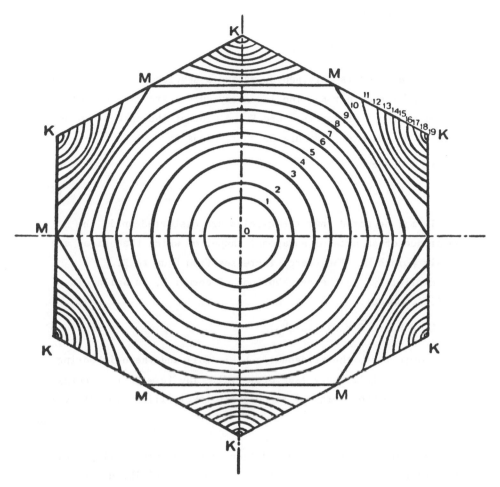

FIG. 20 Isoenergy lines.

In the neighborhood of the center Γ, the components k_1 and k_2 are infinitely small. By using

$$\cos x \simeq 1 - \frac{x^2}{2} \tag{314}$$

and

$$\cos x \cos y = \frac{1}{2}[\cos (x + y) + \cos (x - y)] \tag{315}$$

we obtain the condition

$$k_1^2 + k_2^2 = \text{constant} \tag{316}$$

The same equation is obtained by making a development in the neighborhood of the corner K. The isoenergy curves are thus circular in the neighborhood of points Γ and K. A plot of the curves can be obtained by computer. Figure 20 corresponds to

$$\gamma_0 = 0.9 \text{ eV} \tag{317}$$

$$\gamma_0' = 0.1 \text{ eV} \tag{318}$$

It may be noted that the circles are distorted until the appearance of an area of hexagonal shape, of which the apices are the six points M. The energy increases from Γ up to K.

Explicitly, in the neighborhood of corner K, in Wallace's simplified model,

$$E_- = E_K + \sqrt{3}\ \gamma_0 \pi a\ |\vec{k} - \vec{k}_K| \tag{319}$$

At absolute zero, the hexagonal zone is full and the neighboring zone is completely empty (see Sec. V.C). The term "Fermi energy" is applied to the energy of the highest level filled; in this case,

$$E_{Fermi} = E_K \tag{320}$$

The Fermi-Dirac distribution function gives the probability that a state of energy E will be occupied at thermal equilibrium (Fig. 21):

$$f(E) = \frac{1}{1 + e^{(E-E_K)/kT}} \tag{321}$$

The origin of the energies is subsequently taken at the corner of the Brillouin zone.

The density function of the n(E) states is defined as equal to the number of permissible states enclosed in the crystal volume V for a unit energy interval:

$$n(E)\ dE = \frac{2V}{(2\pi)^3}\ \int_V d\tau_{\vec{k}} \tag{322}$$

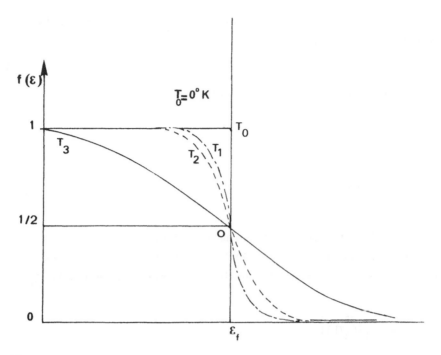

FIG. 21   Fermi-Dirac distribution function.

The factor 2 is due to the spin and the coefficient $(2\pi)^{-3}$ to the fact that the volume evaluated belongs to the reciprocal space.

Equation (322) is thus written [5-8]

$$n(E) = \frac{2V}{(2\pi)^3} \int_{\substack{\text{constant} \\ \text{energy area}}} \frac{dS}{|\vec{\nabla}_k E|} \tag{323}$$

We can perform the calculation in a simple manner. The number of states lying between E and E + dE is, by definition, n(E) dE. How-ever, taking account of the spin, the number of states per unit area in the reciprocal space is

$$u = \frac{2N}{|\vec{b}_1 \wedge \vec{b}_2|} \tag{324}$$

where N is the number of unit cells.

In the neighborhood of corner K, the area lying between two circular isoenergy lines is

$$dS = |2\pi k \; dk| \tag{325}$$

This corresponds to a number of states:

$$n(E) \; dE = u \; dS = \frac{\sqrt{3} \; Na^2}{2\pi} |k \; dk| \tag{326}$$

hence a density $n(E)$ by using (319):

$$n(E) = \frac{N|E - E_K|}{2 \sqrt{3} \; \pi^3 \gamma_0^2} \tag{327}$$

Figure 22 illustrates the foregoing equation.

Comparison of (302) and (305) shows that an equation that accounts for the overlaps is obtained by replacing, in (327), γ_0 by

$$\frac{\gamma_0}{1 + S_{AB}|F_2(\vec{k})|}$$

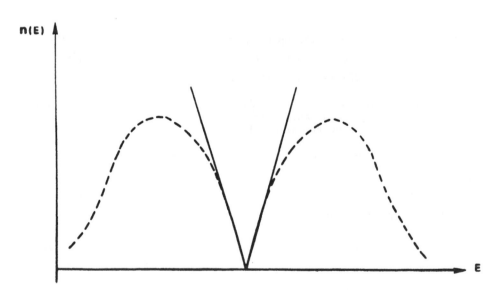

FIG. 22 Density of states as a function of energy: solid line, theoretical curve; dashed line, Coulson's curve (after Refs. 26-28).

At corner K, $|F_2(\vec{k})| = 0$. The density $n(E)$ given by (327) is always valid.

C. Slonczewski-Weiss Model

This is the most widely used model to interpret the physical properties of graphite. It is three-dimensional and assumes orbitals without overlap [condition (303) is satisfied].

Taking account of (290) through (293), the determinant (194) becomes

$$
\begin{vmatrix}
H_{AA} - E & H_{AB} & H_{AC} & H_{AD} \\
H_{AB}^* & H_{BB} - E & H_{AD} & H_{BD} \\
H_{AC} & H_{AD}^* & H_{AA} - E & H_{AB}^* \\
H_{AD}^* & H_{BD}^* & H_{AB} & H_{BB} - E
\end{vmatrix} = 0
\qquad (328)
$$

On the HKH edge of the Brillouin zone (Fig. 19) $k_1 = k_2 = 1/3$, the structural factors have the value

$$
F_1(\vec{k}) = -3 \qquad (310)
$$

$$
F_2(\vec{k}) = \sqrt{3 + F_1(\vec{k})} = F_2^{\lambda}(\vec{k}) = 0 \qquad (329)
$$

This considerably simplifies (328) (see Sec. X.A):

$$
\begin{vmatrix}
H_{AA} - E & 0 & H_{AC} & 0 \\
0 & H_{BB} - E & 0 & 0 \\
H_{AC} & 0 & H_{AA} - E & 0 \\
0 & 0 & 0 & H_{BB} - E
\end{vmatrix} = 0
\qquad (330)
$$

The resolution of (330) gives the solutions

$$
E_1^0 = H_{AA} + H_{AC} \qquad (331)
$$

$$
E_2^0 = H_{AA} - H_{AC} \qquad (332)
$$

$$
E_3^0 = H_{BB} \qquad (333)
$$

$$
E_4^0 = H_{BB} \qquad (334)
$$

The energy E_3^0 is doubly degenerated.

Explicitly, thanks to (285), (287), and (294),

$$E_1^0 = E_0 - \Delta_A + \gamma_5 \cos k_z c + 2\gamma_1 \cos k_z \frac{c}{2} \tag{335}$$

$$E_2^0 = E_0 - \Delta_A + \gamma_5 \cos kz\ c - 2\gamma_1 \cos k_z \frac{c}{2} \tag{336}$$

$$E_3^0 = E_4^0 = E_0 - \Delta_B + \gamma_2 \cos k_z c \tag{337}$$

The corresponding states are

$$\Psi_1 = \frac{1}{\sqrt{2}} (\Psi_A + \Psi_C) \tag{338}$$

$$\Psi_2 = \frac{1}{\sqrt{2}} (\Psi_A - \Psi_C) \tag{339}$$

$$\Psi_3 = \Psi_B \tag{340}$$

$$\Psi_4 = \Psi_D \tag{341}$$

To confirm this, it suffices to show that in this basis of the states, all the nondiagonal elements of (330) are nil and that the secular determinant is written

$$\begin{vmatrix} H_{11} - E & 0 & 0 & 0 \\ 0 & H_{22} - E & 0 & 0 \\ 0 & 0 & H_{33} - E & 0 \\ 0 & 0 & 0 & H_{44} - E \end{vmatrix} = 0 \tag{342}$$

with

$$H_{11} = E_1^0 \tag{343}$$

$$H_{22} = E_2^0 \tag{344}$$

$$H_{33} = E_3^0 \tag{345}$$

$$H_{44} = E_4^0 \tag{346}$$

By convention, let us assume the energy at point H as the origin; that is, as at H: $k_z = \pi/c$, using E_3^0:

$$E_3^0\left(k_z = \frac{\pi}{c}\right) = E_0 - \Delta_B - \gamma_2 = 0 \tag{347}$$

By setting

$$\psi = k_z \frac{c}{2} \tag{348}$$

and

$$\Delta = \Delta_B - \Delta_A + \gamma_2 - \gamma_5 \tag{349}$$

we obtain, taking account of

$$\cos 2\psi = 2 \cos^2 \psi - 1 \tag{350}$$

$$E_1^0 = \Delta + 2\gamma_1 \cos \psi + 2\gamma_5 \cos^2 \psi \tag{351}$$

$$E_2^0 = \Delta - 2\gamma_1 \cos \psi + 2\gamma_5 \cos^2 \psi \tag{352}$$

$$E_3^0 = E_4^0 = 2\gamma_2 \cos^2 \psi \tag{353}$$

Note that if we change the basic vectors and if $\hat{S}$ is the basic change matrix, any matrix $\hat{f}$ is written in the new base:

$$\hat{f}' = \hat{S}^{-1}\hat{f}\hat{S} \tag{354}$$

For example, if the starting base is that of the functions Ψ_A, Ψ_B, Ψ_C, and Ψ_D, and if the new base consists of the vectors Ψ_1, Ψ_2, Ψ_3, and Ψ_4 defined by (338) through (341),

$$\hat{S} = \begin{pmatrix} \frac{1}{\sqrt{2}} & \frac{1}{\sqrt{2}} & 0 & 0 \\ 0 & 0 & 1 & 0 \\ \frac{1}{\sqrt{2}} & -\frac{1}{\sqrt{2}} & 0 & 0 \\ 0 & 0 & 0 & 1 \end{pmatrix} \tag{355}$$

and

$$\hat{S}^{-1} = \begin{pmatrix} \dfrac{1}{\sqrt{2}} & 0 & \dfrac{1}{\sqrt{2}} & 0 \\ \dfrac{1}{\sqrt{2}} & 0 & -\dfrac{1}{\sqrt{2}} & 0 \\ 0 & 1 & 0 & 0 \\ 0 & 0 & 0 & 1 \end{pmatrix} \tag{356}$$

It is easy to confirm that

$$\hat{S}^{-1} \begin{pmatrix} H_{AA} & 0 & H_{AC} & 0 \\ 0 & H_{BB} & 0 & 0 \\ H_{AC} & 0 & H_{AA} & 0 \\ 0 & 0 & 0 & H_{BB} \end{pmatrix} \hat{S} = \begin{pmatrix} H_{AA}+H_{AC} & 0 & 0 & 0 \\ 0 & H_{AA}-H_{AC} & 0 & 0 \\ 0 & 0 & H_{BB} & 0 \\ 0 & 0 & 0 & H_{BB} \end{pmatrix} \tag{357}$$

and hence to demonstrate, that in the new base, the secular determinant is clearly written as in (342).

The variation in energy as a function of k_z is shown in Fig. 23. At H, $k_z = \pi/c$; that is, $\psi = \pi/2$. $\tag{358}$

$$E_1^0 = E_2^0 = \Delta \tag{359}$$

$$E_3^0 = 0 \tag{360}$$

At K, $k_z = 0$ and $\psi = 0$. $\tag{361}$

$$E_1^0 = \Delta + 2\gamma_1 + 2\gamma_5 \tag{362}$$

$$E_2^0 = \Delta - 2\gamma_1 + 2\gamma_5 \tag{363}$$

$$E_3^0 = 2\gamma_2 \tag{364}$$

The bands E_1^0 and E_2^0 can each contain two states per unit cell; the band E_3^0 can contain four states. As it happens, we have four electrons per unit cell to be placed; hence at absolute zero the Fermi level E_F cuts the E_3^0 band (see Fig. 23):

$$2\gamma_2 < E_F < 0 \tag{365}$$

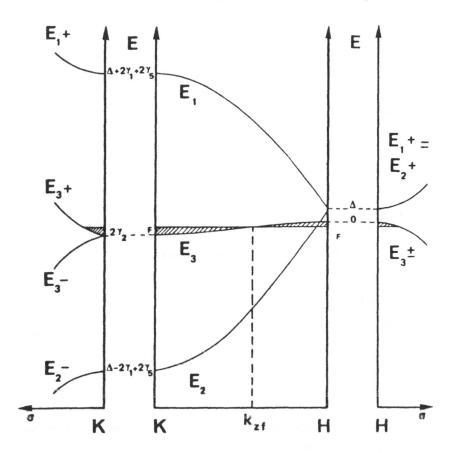

FIG. 23 Variations in energy on the edge of the Brillouin zone.

Let us note k_{z_F}, the value of k_z corresponding to

$$E_F = E_3^U \tag{366}$$

For $k_z < k_{z_F}$, the E_3^0 states are occupied by electrons. For $k_z > k_{z_F}$, the E_3^0 states are occupied by holes.

This localization of charge carriers is conditioned by the sign of γ_2, and a sign change implies a permutation of electrons and holes.

Formulas (351) through (353) give the energy for a point P of the edge HKH defined in reciprocal space by

$$\overrightarrow{\Gamma P} = \overrightarrow{\Gamma K} + \overrightarrow{KP} = \frac{\vec{b}_1 + \vec{b}_2}{3} + k_z \vec{b}_z \tag{367}$$

Let us now examine the energy in the neighborhood of P for a point N (with the same k_z as P), such that, in Fig. 24,

$$\overrightarrow{PN} = \vec{X} \tag{368}$$

The vector $\vec{X}$ forms an angle α with the bisector plane of $(\vec{b}_1, \vec{b}_2)$.

The factors $F_2(\vec{k})$ and $F_2(\vec{k})^*$ are no longer nil but can be evaluated by a first-order development:

$$F_2(\vec{k}) = \sum_B e^{i\vec{k} \cdot \overrightarrow{AB}} = \sum_B e^{i\overrightarrow{\Gamma N} \cdot \overrightarrow{AB}}$$

$$= \sum_B e^{i(\overrightarrow{\Gamma P} + \overrightarrow{PN}) \cdot \overrightarrow{AB}} = \sum_B e^{i\overrightarrow{\Gamma P} \cdot \overrightarrow{AB}} e^{i\vec{X} \cdot \overrightarrow{AB}} \tag{369}$$

Since $\vec{X}$ is small,

$$e^{i\vec{X} \cdot \overrightarrow{AB}} = 1 + i\vec{X} \cdot \overrightarrow{AB} \tag{370}$$

and

$$F_2(\vec{k}) = \sum_B e^{i\overrightarrow{\Gamma P} \cdot \overrightarrow{AB}} + i \sum_B \vec{X} \cdot \overrightarrow{AB} e^{i\overrightarrow{\Gamma P} \cdot \overrightarrow{AB}} \tag{371}$$

The first term corresponds to the structural factor on the edge and is nullified; for the second,

$$\vec{X} = k_1 \vec{b}_1 + k_2 \vec{b}_2 + k_z \vec{b}_z \tag{372}$$

and for the three neighboring atoms,

$$\overrightarrow{AB}_1 = \left(\frac{1}{3}, \frac{2}{3}, 0 \right) \tag{373}$$

$$\overrightarrow{AB}_2 = \left(-\frac{2}{3}, -\frac{1}{3}, 0 \right) \tag{374}$$

$$\overrightarrow{AB}_3 = \left(\frac{1}{3}, -\frac{1}{3}, 0 \right) \tag{375}$$

and

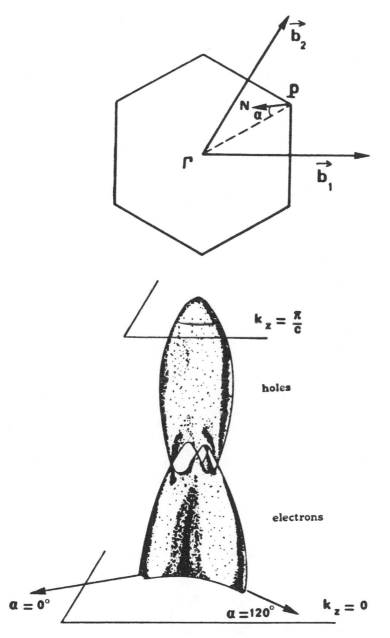

FIG. 24 Fermi surface.

$$F_2(\vec{k}) = 2\pi i \left[\frac{k_1 + 2k_2}{3} e^{\frac{2\pi i}{3}} - \frac{2k_1 + k_2}{3} e^{-\frac{2\pi i}{3}} + \frac{k_1 - k_2}{3} \right]$$

$$= \pi \left[-\sqrt{3}(k_1 + k_2) + i(k_1 - k_2) \right] \tag{376}$$

The components k_1 and k_2 are inferred from the scalar products

$$\vec{X} \cdot \vec{b}_1 = b^2 \left(k_1 + \frac{k_2}{2} \right) = Xb \cos \left(\alpha - \frac{5\pi}{6} \right) = -\frac{Xb}{2}(\sqrt{3} \cos \alpha - \sin \alpha) \tag{377}$$

$$\vec{X} \cdot \vec{b}_2 = b^2 \left(\frac{k_1}{2} + k_2 \right) = Xb \cos \left(\alpha + \frac{5\pi}{6} \right) = \frac{Xb}{2}(\sqrt{3} \cos \alpha + \sin \alpha) \tag{378}$$

Hence,

$$k_1 + k_2 = -\frac{2X}{b\sqrt{3}} \cos \alpha \tag{379}$$

and

$$k_1 - k_2 = \frac{2X}{b} \sin \alpha \tag{380}$$

Thus,

$$F_2(\vec{k}) = \frac{2\pi}{b} X e^{i\alpha} = \frac{a\sqrt{3}}{2} X e^{i\alpha} \tag{381}$$

Setting

$$\sigma = \frac{a\sqrt{3}}{2} X \tag{382}$$

This gives

$$F_2(\vec{k}) = \sigma e^{i\alpha} \tag{383}$$

Let us express the matrix elements H_{ij} in the base of the vectors ψ_i with $i = 1, 2, 3, 4$ defined by (355):

$$(H_{ij}) = \begin{pmatrix} H_{AA} & H_{AB} & H_{AC} & H_{AD} \\ H_{BA} & H_{BB} & H_{BC} & H_{BD} \\ H_{CA} & H_{CB} & H_{CC} & H_{CD} \\ H_{DA} & H_{DB} & H_{DC} & H_{DD} \end{pmatrix} \tag{384}$$

becomes, with the help of (354),

$$(H'_{ij}) = \begin{pmatrix} H_{11} & H_{12} & H_{13} & H_{14} \\ H_{21} & H_{22} & H_{23} & H_{24} \\ H_{31} & H_{32} & H_{33} & H_{34} \\ H_{41} & H_{42} & H_{43} & H_{44} \end{pmatrix} \tag{385}$$

Thus,

$$H_{11} = H_{AA} + H_{AC} = E_1^0 \tag{386}$$

$$H_{12} = H_{21} = 0 \tag{387}$$

$$H_{13} = \frac{1}{\sqrt{2}}(H_{AB} + H_{CB}) = \frac{1}{\sqrt{2}}(2\gamma_4 \cos \psi - \gamma_0)F_2 = (x - y)\sigma e^{i\alpha} \tag{388}$$

$$H_{14} = \frac{1}{\sqrt{2}}(H_{AD} + H_{CD}) = \frac{1}{\sqrt{2}}(2\gamma_4 \cos \psi - \gamma_0)F_2^* = (x - y)\sigma e^{-i\alpha} \tag{389}$$

$$H_{22} = H_{AA} - H_{AC} = E_2^0 \tag{390}$$

$$H_{23} = \frac{1}{\sqrt{2}}(H_{AB} - H_{CB}) = \frac{1}{\sqrt{2}}(\gamma_0 + 2\gamma_4 \cos \psi)F_2 = -(x + y)\sigma e^{i\alpha} \tag{391}$$

$$H_{24} = \frac{1}{\sqrt{2}}(H_{AD} - H_{CD}) = \frac{1}{\sqrt{2}}(\gamma_0 + 2\gamma_4 \cos \psi)F_2^* = (x + y)\sigma e^{-i\alpha} \tag{392}$$

$$H_{34} = H_{BD} = 2\gamma_3 \cos \psi \sigma e^{i\alpha} \tag{393}$$

$$H_{44} = H_{DD} = E_3^0 \tag{394}$$

If

$$x = \sqrt{2}\, \gamma_4 \cos \psi \tag{395}$$

$$y = \frac{\gamma_0}{\sqrt{2}} \tag{396}$$

To find the neighboring states of the E_1^0 and E_2^0 bands and by assuming that γ_3 is sufficiently small, the element H_{34} is ignored in comparison with $E_3^0 - E$ and the development of the secular determinant gives rise to the equation

$$(E_1^0 - E)(E_2^0 - E)(E_3^0 - E)^2 + 2(E_1^0 - E)(E - E_3^0)(x + y)^2\sigma^2$$

$$+ 2(E_2^0 - E)(E - E_3^0)(x - y)^2\sigma^2 + 4(x^2 - y^2)^2\sigma^4 = 0 \tag{397}$$

of which the solutions, taking for origin the value of E_3^0 at point H, are

$$E_{1_\pm} = \frac{1}{2}(E_1 + E_3) \pm \left[\frac{1}{4}(E_1 - E_3)^2 + (\gamma_0 - 2\gamma_4 \cos \psi)^2 \sigma^2\right]^{1/2} \quad (398)$$

$$E_{2_\pm} = \frac{1}{2}(E_2 + E_3) \pm \left[\frac{1}{4}(E_2 - E_3)^2 + (\gamma_0 + 2\gamma_4 \cos \psi)^2 \sigma^2\right]^{1/2} \quad (399)$$

These branches display a symmetry of revolution about HKH and a hyperbolic variation as a function of σ. If σ is small, a development of the root to the first order gives

$$E_{1_\pm} = E_1 + \frac{(\gamma_0 - 2\gamma_4 \cos \psi)^2}{E_1 - E_3} \sigma^2 \quad (400)$$

yielding a parabolic law.

For neighboring levels of E_3^0, this approximation is not valid, but $E_1^0 - E$ can be replaced by

$$E_1^0 - E_3^0 \simeq E_1^0 - E_3$$

or $E_1 - E_3$ and $E_2^0 - E$ by $E_2 - E_3$.

This gives a second-degree equation for which the roots are

$$E_{3_\pm} = E_3 + A\sigma^2 \pm [B^2\sigma^4 + 4B\gamma_3\sigma^3 \cos \psi \cos 3\alpha + 4\sigma^2\gamma_3^2 \cos^2 \psi]^{1/2} \quad (401)$$

with

$$A = \frac{1}{2}\left[\frac{(\gamma_0 - 2\gamma_4 \cos \psi)^2}{E_3 - E_1} + \frac{(\gamma_0 + 2\gamma_4 \cos \psi)^2}{E_3 - E_2}\right] \quad (402)$$

$$B = \frac{1}{2}\left[\frac{(\gamma_0 - 2\gamma_4 \cos \psi)^2}{E_3 - E_1} - \frac{(\gamma_0 + 2\gamma_4 \cos \psi)^2}{E_3 - E_2}\right] \quad (403)$$

The degeneration of level E_3 is obviated except in the neighborhood of point H; that is, $k_z = \pi/c$ where $\cos \psi$ and B are nil (Fig. 23).

The presence of "$\cos 3\alpha$" implies a ternary symmetry about HKH conditioned by γ_3 and $\cos \psi$: if $\gamma_3 = 0$ or $k_z = \pi/c$, a symmetry of revolution is obtained. This explains the form of the Fermi surface, the locus of points of energy E_F, of which the cross section is circular in the neighborhood of H and exhibits a ternary distortion, "trigonal warping," in the neighborhood of K (Fig. 24).

GRAPHITE: MAGNETIC HAMILTONIAN

XI. HAMILTONIAN DESCRIPTION OF SYSTEMS [40]

A. Review of Point Mechanics

In an orthonormed trihedron, with origin 0, a particle of mass m is identified by the vector

$$\vec{r} = \overrightarrow{OM} = \vec{e}_1 x_1 + \vec{e}_2 x_2 + \vec{e}_3 x_3 \tag{404}$$

$\vec{e}_1$, $\vec{e}_2$, and $\vec{e}_3$ are the unitary vectors of the referential axes. This particle has a velocity $\vec{v}$, measured in relation to 0, given by

$$\vec{v} = \frac{d\vec{r}}{dt} \tag{405}$$

and a momentum

$$\vec{p} = m\vec{v} \tag{406}$$

The kinetic moment $\vec{L}_0$, calculated at 0, is defined by

$$\vec{L}_0 = \overrightarrow{OM} \wedge \vec{p} \tag{407}$$

Because of its movement, the particle acquires a kinetic energy

$$T = \frac{1}{2} mv^2 = \frac{p^2}{2m} \tag{408}$$

According to d'Alembert's principle,

$$\vec{F} = \frac{d\vec{p}}{dt} \tag{409}$$

Except in relativistic cases, this is written for a constant mass system

$$\vec{F} = m\vec{\gamma} \tag{410}$$

and one speaks of the fundamental theorem of dynamics.

The vector $\vec{\gamma}$ is the acceleration to which the particle is subjected:

$$\vec{\gamma} = \frac{d\vec{v}}{dt} \tag{411}$$

By definition, $\vec{F}$ is the force exerted on the particle of mass m.

By deriving Eq. (4) in relation to time, the so-called kinetic moment theorem is obtained:

$$\frac{d\vec{L}_0}{dt} = \vec{M}_0 \tag{412}$$

where $\vec{M}_0$ is the moment, calculated at 0, of the forces applied to the particle:

$$\vec{M}_0 = \vec{r} \wedge \vec{F} \tag{413}$$

If the force $\vec{F}$ is placed in the form $\vec{F} = \vec{F}(\vec{r})$, one is in the presence of a force field. These forces are called conservative, together with the system, if the work of these forces between two points A and B depends only on the starting and arrival points, not on the path followed from A to B. Mathematically, this is expressed by the equation

$$W_{AB} = \int_A^B \vec{F} \cdot d\vec{r} = V(A) - V(B) \tag{414}$$

where V is the potential energy of the particle. Equation (414) is equivalent to

$$\oint \vec{F} \cdot d\vec{r} = 0 \tag{415}$$

and to

$$\vec{F} = -\vec{\nabla}V = -\overrightarrow{\text{grad}}\ V = -\left(\vec{e}_1 \frac{\partial V}{\partial x_1} + \vec{e}_2 \frac{\partial V}{\partial x_2} + \vec{e}_3 \frac{\partial V}{\partial x_3}\right) \tag{416}$$

Note that by definition the gradient of a point function $f = f(x_1, x_2, x_3)$ is given by

$$df = \overrightarrow{\text{grad}}\ f \cdot d\vec{r} \tag{417}$$

For conservative systems the sum of the quantities T and V is a constant of movement E called the total mechanical energy:

$$E = T + V \tag{418}$$

the kinetic energy T depends exclusively on the velocity, and the potential energy V depends only on the position.

Equations (406) through (408) and (418) are easily generalized to a set of particles, with all the magnitudes defined above additive.

It is customary for the different derivatives to be noted as follows.

Derivative in relation to time t:

$$\dot{x} = \frac{dx}{dt}, \qquad \ddot{x} = \frac{d^2x}{dt^2}, \qquad \cdots \qquad (419)$$

Derivative in relation to a variable other than time, for example, z:

$$x' = \frac{dx}{dz}, \qquad x'' = \frac{d^2x}{dz^2} \qquad (420)$$

B. Lagrangian Formalism

1. Generalized Coordinates

If we consider a system of N particles without interaction, the trajectories of the particles are independent. Each one is identified by three parametric equations of the type

$$x_i = x_i(t) \qquad i = 1, 2, 3, \ldots, 3N \qquad (421)$$

A total of 3N equations needs to be written. The 3N coordinates x_i are independent and serve to construct a hyperspace with 3N dimensions called the configuration space, each coordinate characterizing one degree of freedom of the system.

By definition of x_i, any linear equation of the type

$$\lambda_1 x_1 + \lambda_2 x_2 + \cdots + \lambda_{3N} x_{3N} = 0 \qquad (422)$$

can only be confirmed if all the λ_i are zero. In the case of interactions between the particles of the system, characterized by the holonomic stress equations, that is, equations relating different x_i (without involving $\dot{x}_i$), the latter possesses d degrees of freedom:

$$d = 3N - \ell \qquad (423)$$

and consequently d independent variables, called generalized coordinates and noted q_k; the configuration space is a hyperspace of
dimension d.

By expressing the x_i coordinates as a function of the variables
q_k, t being considered as a parameter,

$$x_i = x_i(q_1, q_2, \cdots, q_d; t) \tag{424}$$

These are the so-called equations of passage.

2. *Generalized Velocities*

By definition,

$$\dot{x}_i = \frac{dx_i}{dt} \tag{425}$$

and, from the properties of the total differentials with (21),

$$\dot{x}_i = \frac{\partial x_i}{\partial t} + \sum_{j=1}^{d} \frac{\partial x_i}{\partial q_j} \dot{q}_j \tag{426}$$

it is easily inferred that

$$\frac{\partial \dot{x}_i}{\partial \dot{q}_j} = \frac{\partial x_i}{\partial q_j} \tag{427}$$

and that the virtual displacement δx_i is given by

$$\delta x_i = \sum_{j=1}^{d} \frac{\partial x_i}{\partial q_j} \delta q_j \tag{428}$$

3. *Generalized Forces*

The N particles of the system are subjected to forces $\vec{F}_i$ (i = 1, 2,
3, ..., N), and during a virtual displacement $\vec{\delta r}_i$ of the ith particle, the work performed by force $\vec{F}_i$ is written

$$\delta W_i = \vec{F}_i \cdot \vec{\delta r}_i \tag{429}$$

For all the particles, the virtual work W is obtained by

$$\delta W = \sum_{i=1}^{N} \vec{F}_i \cdot \vec{\delta r}_i \tag{430}$$

In accordance with the generalized coordinates, we can replace (430) by

$$\delta W = \sum_{k=1}^{d} F_k \, \delta q_k \tag{431}$$

where, by definition, F_k is the generalized force, so that with the help of (426):

$$\sum_i \vec{F}_i \cdot \delta \vec{r}_i = \sum_{i,k} \vec{F}_i \cdot \frac{\partial \vec{r}_i}{\partial q_k} \, \delta q_k \tag{432}$$

$$F_k = \sum_{i=1}^{N} \vec{F}_i \cdot \frac{\partial \vec{r}_i}{\partial q_k} \tag{433}$$

If the different external forces $\vec{F}_i$ applied to the particles derive from a potential,

$$\vec{F}_i = -\vec{\nabla}_i V_i = -\left(\vec{e}_1 \frac{\partial V_i}{\partial x_{i_1}} + \vec{e}_2 \frac{\partial V_i}{\partial x_{i_2}} + \vec{e}_3 \frac{\partial V_i}{\partial x_{i_3}} \right) \tag{434}$$

the potential energy is an additive magnitude and the system is characterized by a total potential energy

$$V = \sum_{i=1}^{N} V_i \tag{435}$$

By rewriting (433), we have

$$F_k = - \sum_{i=1}^{N} \vec{\nabla}_i V_i \cdot \frac{\partial \vec{r}_i}{\partial q_k} \tag{436}$$

By definition of the gradient of a function,

$$\vec{\nabla}_i V_i \, \partial \vec{r}_i = \partial V_i \tag{437}$$

and

$$F_k = - \sum_{i=1}^{N} \frac{\partial V_i}{\partial q_k} = - \frac{\partial V}{\partial q_k} \tag{438}$$

Equation (438) is equivalent to (434) and shows that the generalized force F_k derives from the same potential as $\vec{F}_i$, but in the set of generalized coordinates.

4. *Generalized Lagrangian Equation*

According to d'Alembert's principle and the theorem of virtual work,

$$\sum_{i=1}^{N} \left(\vec{F}_i - \frac{d\vec{p}_i}{dt} \right) \cdot \delta \vec{r}_i = 0 \tag{439}$$

and with (428),

$$\sum_{i=1}^{N} \sum_{j=1}^{d} \left(\vec{F}_i - \frac{d\vec{p}_i}{dt} \right) \cdot \frac{\partial \vec{r}_i}{\partial q_j} \delta q_j = 0 \tag{440}$$

or

$$\sum_{i,j} m_i \ddot{\vec{r}}_i \cdot \frac{\partial \vec{r}_i}{\partial q_j} \delta q_j - \sum_{i,j} \frac{d\vec{p}_i}{dt} \cdot \frac{\partial \vec{r}_i}{\partial q_j} \delta q_j = 0 \tag{441}$$

But,

$$\sum_i \frac{d}{dt} \left(m_i \dot{\vec{r}}_i \cdot \frac{\partial \vec{r}_i}{\partial q_j} \right) = \sum_i m_i \ddot{\vec{r}}_i \cdot \frac{\partial \vec{r}_i}{\partial q_j} + \sum_i m_i \dot{\vec{r}}_i \cdot \frac{\partial \vec{v}_i}{\partial q_j} \tag{442}$$

and, with (427),

$$\sum_i m_i \ddot{\vec{r}}_i \cdot \frac{\partial \vec{r}_i}{\partial q_j} = \sum_i \frac{d}{dt} \left(m_i \vec{v}_i \cdot \frac{\partial \vec{v}_i}{\partial \dot{q}_j} \right) - \sum_i m_i \vec{v}_i \cdot \frac{\partial \vec{v}_i}{\partial q_j} \tag{443}$$

By rewriting (441),

$$\sum_{i,j} \left[\frac{d}{dt} \left(m_i \vec{v}_i \cdot \frac{\partial \vec{v}_i}{\partial \dot{q}_j} \right) - m_i \vec{v}_i \cdot \frac{\partial \vec{v}_i}{\partial q_j} - \vec{p}_i \cdot \frac{\partial \vec{r}_i}{\partial q_j} \right] \delta q_j = 0 \tag{444}$$

Note that

$$\sum_i \vec{p}_i \cdot \frac{\partial \vec{r}_i}{\partial q_j} = \sum_i \vec{F}_i \cdot \frac{\partial \vec{r}_i}{\partial q_j} = F_j \qquad \text{see Eq. (433)}$$

Hence,

$$\sum_j \left\{ \sum_i \left[\frac{d}{dt} \left(m_i \vec{v}_i \cdot \frac{\partial \vec{v}_i}{\partial \dot{q}_j} \right) - m_i \vec{v}_i \cdot \frac{\partial \vec{v}_i}{\partial q_j} \right] - F_j \right\} \delta q_j = 0 \tag{445}$$

Since the δq_j are independent, it is necessary for all the coefficients of the δq_j in (445) to be zero; that is,

$$\sum_i \left\{ \frac{d}{dt} \left[m_i \vec{v}_i \cdot \frac{\partial \vec{v}_i}{\partial \dot{q}_j} \right] - m_i \vec{v}_i \cdot \frac{\partial \vec{v}_i}{\partial q_j} \right\} = F_j \tag{446}$$

The total kinetic energy T of the system is given by

$$T = \frac{1}{2} \sum_i m_i v_i^2 \qquad \text{and} \qquad \frac{\partial T}{\partial q_j} = \sum_i m_i \vec{v}_i \cdot \frac{\partial \vec{v}_i}{\partial q_j} \tag{447}$$

$$\frac{\partial T}{\partial \dot{q}_j} = \sum_i m_i \vec{v}_i \cdot \frac{\partial \vec{v}_i}{\partial \dot{q}_j} \tag{448}$$

Equation (446) is transformed into

$$\frac{d}{dt} \left(\frac{\partial T}{\partial \dot{q}_j} \right) - \frac{\partial T}{\partial q_j} = F_j \tag{449}$$

This equation is valid irrespective of the types of force present, conservative or not.

If F_j derives from the potential energy (438), we have

$$\frac{d}{dt} \left(\frac{\partial T}{\partial \dot{q}_j} \right) - \frac{\partial}{\partial q_j} (T - V) = 0 \tag{450}$$

If we define the Lagrangian function

$$L(q_k, \dot{q}_k, t) = T(\dot{q}_k) - V(q_k) \tag{451}$$

Thus (450) is the generalized Lagrangian equation

$$\frac{d}{dt} \left(\frac{\partial L}{\partial \dot{q}_k} \right) - \frac{\partial L}{\partial q_k} = 0 \tag{452}$$

By extension of the notion of momentum, the generalized moments are defined by

$$p_k = \frac{\partial L}{\partial \dot{q}_k} \tag{453}$$

In fact, for a motion occurring along the O_{x_1} axis,

$$T = \frac{1}{2} m\dot{x}_1^2 \qquad \text{and} \qquad \frac{\partial T}{\partial \dot{x}_1} = \frac{\partial L}{\partial \dot{x}_1} = m\dot{x}_1 = p_{x_1}$$

Thus from (452),

$$\dot{p}_k = \frac{\partial L}{\partial q_k} \tag{454}$$

5. *Generalized Kinetic Energy*

For all the particles of a given physical system of which the kinetic energy is an additive magnitude,

$$T = \frac{1}{2} \sum_i m_i \dot{r}_i^2 \tag{455}$$

since

$$\vec{r}_i = \frac{\partial \vec{r}_i}{\partial t} + \sum_j \frac{\partial \vec{r}_i}{\partial q_j} \dot{q}_j \tag{426}$$

Hence,

$$T = \frac{1}{2} \sum_i m_i \left(\frac{\partial \vec{r}_i}{\partial t} + \sum_j \frac{\partial \vec{r}_i}{\partial q_j} \dot{q}_j \right) \cdot \left(\frac{\partial \vec{r}_i}{\partial t} + \sum_k \frac{\partial \vec{r}_i}{\partial q_k} \dot{q}_k \right)$$

$$= \alpha + \sum_j \alpha_j \dot{q}_j + \sum_{j,k} \alpha_{jk} \dot{q}_j \dot{q}_k \tag{456}$$

with

$$\alpha = \frac{1}{2} \sum_i m_i \left(\frac{\partial \vec{r}_i}{\partial t} \right)^2$$

$$\alpha_j = \sum_i m_i \frac{\partial \vec{r}_i}{\partial t} \cdot \frac{\partial \vec{r}_i}{\partial q_j} \tag{457}$$

$$\alpha_{jk} = \frac{1}{2} \sum_i m_i \frac{\partial \vec{r}_i}{\partial q_j} \frac{\partial \vec{r}_i}{\partial q_k}$$

Equation (456) may be reduced. If the position vector $\vec{r}_i = \overrightarrow{OM}_i$ is not an explicit function of time,

$$\frac{\partial \vec{r}_i}{\partial t} = \vec{0}$$

giving rise to

$$\alpha = \alpha_j = 0 \qquad \text{and} \qquad T = \sum_{j,k} \alpha_{jk} \dot{q}_j \dot{q}_k \tag{458}$$

C. Legendre's Transformation [41]

1. *Phase Space*

The term "phase space" is applied to the hyperspace with 2d dimensions in which a point is identified by the independent variables $q_1, q_2, \ldots, q_d; p_1, p_2, \ldots, p_d$.

2. *Tangent Plane Equation*

Let us find the equation of the tangent plane to a surface Σ defined by

$$F(x, y, z) = C \tag{459}$$

where C is a constant.

Let $P(x, y, z)$ be a point of the surface Σ and $Q(\overline{x}, \overline{y}, \overline{z})$ a point of the tangent plane to Σ at P (Fig. 25). Since

$$dF = 0 \qquad \text{and} \qquad dF = \vec{\nabla}F \cdot \vec{dr} \tag{460}$$

the vector $\vec{\nabla}F$ is perpendicular to $\vec{dr}$, that is, to surface Σ, hence to vector $\vec{PQ}$.

Now let us clarify (460):

$$(\overline{x} - x)\frac{\partial F}{\partial x} + (\overline{y} - y)\frac{\partial F}{\partial y} + (\overline{z} - z)\frac{\partial F}{\partial z} = 0 \tag{461}$$

Consider the function

$$z = z(x, y) \tag{462}$$

where x and y are two independent variables. We can assume

$$F(x, y, z) = z - z(x, y) = 0 \tag{463}$$

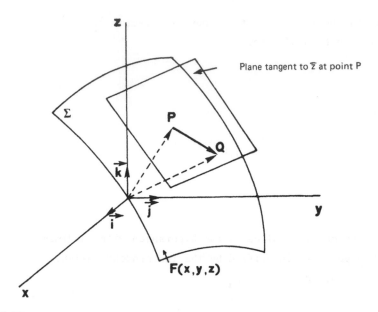

FIG. 25   Tangent plane to a surface.

According to (461), we have

$$-(\overline{x} - x)\frac{\partial z}{\partial x} - (\overline{y} - y)\frac{\partial z}{\partial y} + (\overline{z} - z) = 0$$

or

$$\overline{z} - a_1\overline{x} - a_2\overline{y} + c = 0 \qquad\qquad (464)$$

with

$$a_1 = \frac{\partial z}{\partial x}$$

$$a_2 = \frac{\partial z}{\partial y} \qquad\qquad (465)$$

$$c = x\frac{\partial z}{\partial x} + y\frac{\partial z}{\partial y} - z = a_1 x + a_2 y - z$$

c is a function of a_1 and a_2:

$$x = \frac{\partial c}{\partial a_1}$$

$$\qquad\qquad (466)$$

$$y = \frac{c}{a_2}$$

The term "Legendre's transformation" is applied to the equation

$$z(x, y) + c(a_1, a_2) = a_1 x + a_2 y \tag{467}$$

where x and y are given by Eqs. (466). The transformation is only valid if a bijection can be made in the analytic area between the coordinates of the points on the surface and those of the tangent planes.

Analytically it is necessary for the Jacobean

$$\frac{\partial(a_1, a_2)}{\partial(x_1, x_2)} = \begin{vmatrix} \dfrac{\partial a_1}{\partial x_1} & \dfrac{\partial a_1}{\partial x_2} \\[2ex] \dfrac{\partial a_2}{\partial x_1} & \dfrac{\partial a_2}{\partial x_2} \end{vmatrix} \neq 0 \tag{468}$$

D. Hamiltonian Function

1. *Definition*

Bohr-Sommerfeld's rules of quantification,

$$\oint p_k \, dq_k = n_k \hbar \tag{469}$$

show that the phase space is adapted to the study of quantum systems. Legendre's transformation allows us to go from a representation q_k, $\dot{q}_k$; t to this space q_k, p_k; t. If we take the Lagrangian function for function z, Eq. (467) is written

$$L(q_k, \dot{q}_k; t) + c(q_k, p_k; t) = p_k \dot{q}_k \tag{470}$$

By generalizing (470) and noting the quantity $c(q_k, p_k; t)$ as the Hamiltonian function H, we have

$$H(q_k, p_k; t) = \sum_{k=1}^{d} p_k \dot{q}_k - L(q_k, \dot{q}_k; t) \tag{471}$$

2. *Canonical Equations*

Let us differentiate the Hamiltonian function

$$dH = \sum_k p_k d\dot{q}_k + \sum_k \dot{q}_k dp_k - dL \tag{472}$$

hence,

$$dL = \frac{\partial L}{\partial t} dt + \sum_k \frac{\partial L}{\partial q_k} dq_k + \sum_k \frac{\partial L}{\partial \dot{q}_k} d\dot{q}_k \qquad (473)$$

and, with the help of (453) and (454),

$$dH = -\sum_k \dot{p}_k dq_k + \sum_k \dot{q}_k dp_k - \frac{\partial L}{\partial t} dt \qquad (474)$$

From (474), we obtain

$$\frac{\partial H}{\partial t} = -\frac{\partial L}{\partial t}$$

$$\dot{p}_k = -\frac{\partial H}{\partial q_k} \qquad (475)$$

$$\dot{q}_k = \frac{\partial H}{\partial p_k}$$

These are the so-called canonical or Hamiltonian equations.

3.   *Conservative Systems*

If the Lagrangian function does not depend explicitly on time, (474) gives

$$\frac{dH}{dt} = 0 \qquad (476)$$

The Hamiltonian function $H(q_k, p_k)$ is then a constant of motion.
By definition of the functions L (451) and H [Eqs. (471) and (453)],

$$H = \sum_k p_k \dot{q}_k - T + V = \sum_k \frac{\partial L}{\partial \dot{q}_k} \dot{q}_k - T + V \qquad (477)$$

In certain conditions,

$$T = \sum_{i,j} \alpha_{ij} \dot{q}_i \dot{q}_j \qquad (458)$$

that is,

$$\frac{\partial T}{\partial \dot{q}_k} = 2 \sum_j \alpha_{kj} \dot{q}_j \qquad (478)$$

Since,

$$\frac{\partial T}{\partial \dot{q}_k} = \frac{\partial L}{\partial \dot{q}_k}$$

Eq. (477) is written

$$H = 2 \sum_{j,k} \alpha_{kj} \dot{q}_j \dot{q}_k - T + V$$

or even, owing to (458),

$$H = T + V \tag{479}$$

For any conservative system,

$$H = \sum_k p_k \dot{q}_k - L(q_k, \dot{q}_k)$$

But for (479) to be verified, it is also necessary that the equations of passage from $\vec{r}_i$ to q_k do not contain time explicitly. The Hamiltonian function is then the total mechanical energy of the system. We can write $H = H(q_k, p_k)$, where the moments p_k are evaluated by means of (453) and assuming $L(q_k, \dot{q}_k) = T - V$.

XII. MAGNETIC HAMILTONIAN $\hat{H}_m$ [42]

A. Gauge Employed

The magnetic vector $\vec{B}$ is connected to the vector potential $\vec{A}$ by the formula

$$\vec{B} = \vec{\nabla} \wedge \vec{A} = \begin{vmatrix} \vec{i} & \vec{j} & \vec{k} \\ \frac{\partial}{\partial x} & \frac{\partial}{\partial y} & \frac{\partial}{\partial z} \\ A_x & A_y & A_z \end{vmatrix} \tag{480}$$

The components of the induction vector are thus

$$B_x = \frac{\partial A_z}{\partial y} - \frac{\partial A_y}{\partial z} \qquad B_y = \frac{\partial A_x}{\partial z} - \frac{\partial A_z}{\partial x} \qquad B_z = \frac{\partial A_y}{\partial x} - \frac{\partial A_y}{\partial y} \tag{481}$$

If one selects

$$A_y = A_z = 0 \tag{482}$$

then

$$B_x = 0 \tag{483}$$

$$B_y = \frac{\partial A_x}{\partial z} \tag{484}$$

$$B_z = -\frac{\partial A_x}{\partial y} \tag{485}$$

and, if in addition the axis O_z is selected parallel to vector $\vec{B}$,

$$B_x = B_y = 0 \tag{486}$$

and

$$B_z = -\frac{\partial A_x}{\partial y} \tag{487}$$

hence,

$$A_x = -By \tag{488}$$

considering the case of a constant magnetic field.

B. Writing of $\hat{H}_m$

1. *Generalized Potential Energy*

Now let us consider a particle of mass m with charge q subjected to a force that derives from the potential energy W such that

$$\vec{F} = -\overrightarrow{\text{grad}}\ W \tag{489}$$

and placed in a region of space in which the magnetic induction $\vec{B}$ is zero. Obviously, we must have the equations

$$\vec{F} = q\vec{E} \tag{490}$$

and

$$W = qV \tag{491}$$

where $\vec{E}$ is the electric field vector and V the scalar potential. Equations (490) and (491) constitute the definitions of these two magnitudes. Hence by comparison between (489) and (490),

$$\vec{E} = -\overrightarrow{\text{grad}} \, V \tag{492}$$

but also,

$$\vec{\nabla} \wedge \vec{E} = \begin{vmatrix} \vec{i} & \vec{j} & \vec{k} \\ \dfrac{\partial}{\partial x} & \dfrac{\partial}{\partial y} & \dfrac{\partial}{\partial z} \\ -\dfrac{\partial V}{\partial x} & -\dfrac{\partial V}{\partial y} & -\dfrac{\partial V}{\partial z} \end{vmatrix} \tag{493}$$

If we assume $\vec{B} \neq \vec{0}$, Eq. (490) is replaced by

$$\vec{F} = q[\vec{E} + \vec{v} \wedge \vec{B}] \tag{494}$$

Laplacian force, and (493) is written

$$\vec{\nabla} \wedge \vec{E} = -\frac{\partial \vec{B}}{\partial t} \tag{495}$$

one of the "Maxwellian equations."

This leads to the substitution for $\vec{E}$ given by (492):

$$\vec{E} = -\vec{\nabla}V - \frac{\partial \vec{A}}{\partial t} \tag{496}$$

One can therefore write

$$\vec{F} = q\left[-\vec{\nabla}V - \frac{\partial \vec{A}}{\partial t} + \vec{v} \wedge (\vec{\nabla} \wedge \vec{A})\right] \tag{497}$$

By projecting on the 0_x axis,

$$m\ddot{x} = q\left[-\frac{\partial V}{\partial x} - \frac{\partial A_x}{\partial t} + \dot{y}\left(\frac{\partial A_y}{\partial x} - \frac{\partial A_x}{\partial y}\right) - \dot{z}\left(\frac{\partial A_x}{\partial z} - \frac{\partial A_z}{\partial x}\right)\right] \tag{498}$$

This equation can also be derived from the generalized Lagrangian Eq. (452):

$$\frac{d}{dt}\left(\frac{\partial L}{\partial \dot{q}_k}\right) - \frac{\partial L}{\partial q_k} = 0$$

provided that a generalized potential energy $q\dot{\vec{r}}\cdot\vec{A} - qV$ is taken in L; that is,

$$L = \frac{1}{2}m\dot{r}^2 + q\dot{\vec{r}}\cdot\vec{A} - qV \tag{499}$$

Since

$$p_k = \frac{\partial L}{\partial \dot{q}_k} \tag{453}$$

we have

$$p_x = m\dot{x} + qA_x \tag{500}$$

which is easily generalized to

$$\vec{p} = m\dot{\vec{r}} + q\vec{A} \tag{501}$$

This equation helps to calculate the Hamiltonian function, so that we can write [Eq. (471)]

$$H = \vec{p}\cdot\dot{\vec{r}} - L = \frac{1}{2}m\dot{r}^2 + qV$$

or

$$H = \frac{1}{2m}(\vec{p} - q\vec{A})^2 + qV \tag{502}$$

2. *Magnetic Hamiltonian Operator*

However, in the wave function space, the functional expression of the operator $\hat{p}_x$ is

$$\hat{p}_x \equiv -i\hbar\frac{\hat{\partial}}{\partial x} \tag{503}$$

and that of the commutator,

$$[\hat{x}, \hat{p}_x] = \hat{x}\hat{p}_x - \hat{p}_x\hat{x} = i\hbar \tag{504}$$

One can thus evaluate the quantity

$$[\hat{x}^2, \hat{p}_x] = \hat{x}[\hat{x}, \hat{p}_x] + [\hat{x}, \hat{p}_x]\hat{x} = 2i\hbar\hat{x} \tag{505}$$

By recurrence it is easily demonstrated that

$$[\hat{x}^n, \hat{p}_x] = ni\hbar\hat{x}^{n-1} = i\hbar\frac{\partial}{\partial x}x^n \tag{506}$$

Any operator $\hat{O}(\hat{x})$ can be developed in series of powers of $\hat{x}$ and

$$[\hat{O}(\hat{x}), \hat{p}_x] = i\hbar\frac{\partial}{\partial x}\hat{O}(\hat{x})$$

Furthermore, if the space $\{|\vec{r}>\}$ is the tensorial product space

$$|\vec{r}> = |x> \otimes |y> \otimes |z>$$

and the commutation rules in this space are

$$[\hat{r}, \hat{p}_x] = i\hbar$$

and

$$[\hat{O}(\hat{r}), \hat{p}_x] = i\hbar\frac{\partial}{\partial x}\hat{O}(\hat{r}) \tag{507}$$

For further details we recommend reading Ref. 43. The identity (502) is written

$$H = \frac{p^2}{2m} + \frac{q^2A^2}{2m} + qV - \frac{q}{2m}(\vec{p}\cdot\vec{A} + \vec{A}\cdot\vec{p}) \tag{508}$$

and, thanks to (507), in the form

$$[\hat{\vec{A}}, \hat{\vec{p}}] = i\hbar\frac{\partial}{\partial x}\vec{A} \tag{509}$$

that is,

$$\vec{p}\cdot\vec{A} = \vec{A}\cdot\vec{p} - i\hbar\vec{\nabla}\cdot\vec{A} \tag{510}$$

and gives

$$H = \frac{p^2}{2m} + \frac{q^2A^2}{2m} + qV - \frac{q}{m}\vec{A}\cdot\vec{p} + \frac{i\hbar}{2m}q\vec{\nabla}\cdot\vec{A} \tag{511}$$

Hence, by means of Ehrenfest's theorems, by setting

$$\hat{H} = \hat{H}_0 + \hat{H}_m \tag{512}$$

the operators

$$\hat{H}_0 = \frac{\hat{p}^2}{2m} + q\hat{V} = -\frac{\hbar^2}{2m}\hat{\nabla}^2 + q\hat{V} \tag{513}$$

and

$$\hat{H}_m = \frac{q^2 A^2}{2m} + \frac{i\hbar q}{2m}(2\vec{A}\cdot\vec{\nabla} + \vec{\nabla}\cdot\vec{A}) \tag{514}$$

The latter is the magnetic Hamiltonian. It is easy to verify the equation

$$\vec{\nabla}\cdot(\vec{A}\Psi) = (\vec{\nabla}\cdot\vec{A})\Psi + \vec{A}\cdot\vec{\nabla}\Psi \tag{515}$$

The vector potential $\vec{A}$ is such that

$$\vec{\nabla}\cdot\vec{A} = 0 \tag{516}$$

where

$$\vec{\nabla}\cdot\vec{A} = \frac{\partial}{\partial x}A_x + \frac{\partial}{\partial y}A_y + \frac{\partial}{\partial z}A_z$$

with $A_x = -By$ [Eq. (488)] and $A_y = A_z = 0$ [Eq. (482)]; hence,

$$\hat{H}_m = \frac{q^2}{2m}A^2 + i\hbar\frac{q}{m}\vec{A}\cdot\vec{\nabla} \tag{517}$$

Equations (488) and (503) serve to write the equation

$$\hat{H}_m = \frac{q^2}{2m}B^2\hat{y}^2 + \frac{q}{m}B\hat{y}\hat{p}_x \tag{518}$$

which is the starting point of the calculations of Luttinger and Kohn [44].

XIII. EQUATION OF EFFECTIVE MASS

The calculations we intend to develop constitute an adaptation of the theory discussed in Ref. 44. The problem was raised when Ganguli and Krishnan [45] and Hove [46], followed by McClure [14,47], attempted to explain the diamagnetism of graphite. The model due to Landau-Peierls [48,49] offers a theoretical numerical value of the suscepti-bility of conduction electrons about 40 times weaker than the pre-dicted value. Adams [50] showed that the formulas employed failed to apply to the case of neighboring energy bands. The band-to-band transitions induced by a constant external magnetic field intervene too strongly in the calculation of diamagnetic susceptibility.

For graphite, at point H (Fig. 19) situated on the edge of the Brillouin zone, the valence and conduction bands are degenerated (Fig. 23). The minimum energy is not located at the center $\Gamma(\vec{k} = \vec{0})$ but at the corner K of the zone.

A. Choice of the Basis of Eigenfunctions

1. Degenerated Eigenvalue [17,18]

Consider a linear, hermitic operator $\hat{O}$. Let us note $|\Psi_i>$, the eigenvectors of $\hat{O}$ associated with the eigenvalues O_i:

$$\hat{O}|\psi_i> = O_i|\psi_i> \tag{519}$$

If at least two eigenvectors of $\hat{O}$ exist, linearly independent, for the same eigenvalue O_i, the latter is said to be degenerate.

If O_i is degenerated by order d, it corresponds to d independent kets $|\psi_i^j>$

$$j = 1, 2, 3, \ldots, d$$

such that

$$\hat{O}|\psi_i^j> = O_i|\psi_i^j> \tag{520}$$

irrespective of j.

But in this case any ket $|\Psi>$ of the form

$$|\Psi> = \sum_{j=1}^{d} c_j|\psi_i^j> \tag{521}$$

is an eigenvector of $\hat{O}$ for the eigenvalue O_i

$$\hat{O}|\Psi> = \sum_{j=1}^{d} c_j O_i|\psi_i^j> = O_i|\Psi> \tag{522}$$

We can state that the total of the eigenkets of $\hat{O}$ associated with the same eigenvalue O_i constitutes a vectorial space of dimension d called the subeigenspace of the eigenvalue O_i.

2. Orthogonality

The base selected to analyze graphite is that of Bloch's functions, which in the case of degeneration can be written in the form

$$\langle \vec{r} | \phi_{n,\vec{k}}^{j} \rangle = e^{i\vec{k}\cdot\vec{r}} \; \phi_{n}^{j}(\vec{r}) \tag{523}$$

where n is the band number and j = 1, 2, 3, . . ., d with

$$\phi_{n}^{j}(\vec{r} + \vec{T}) = \phi_{n}^{j}(\vec{r}) \tag{524}$$

if

$$\vec{T} = n_{1}\vec{a}_{1} + n_{2}\vec{a}_{2} + n_{3}\vec{a}_{3} \tag{525}$$

where $\vec{a}_1$, $\vec{a}_2$, and $\vec{a}_3$ are the base vectors of the space lattice and n_1, n_2, and n_3 are whole numbers. The form of these wave functions has been examined in Secs. VI-X.

The orthogonality equation of the eigenvectors becomes

$$\langle \phi_{n}^{j} | \phi_{n}^{j'} \rangle = \delta_{jj'} \tag{526}$$

and

$$\langle \phi_{n,\vec{k}}^{j} | \phi_{n',\vec{k}'}^{j'} \rangle = \delta_{jj'} \; \delta_{nn'} \; \delta(\vec{k}' - \vec{k}) \tag{527}$$

In this orthonormed base, any function $\Psi(\vec{r})$ may be developed in series (superposition principle):

$$\Psi(\vec{r}) = \sum_{\substack{j=1 \\ n}}^{d} \sum \int d^{3}k \; a_{n,j}(\vec{k}) \phi_{n,\vec{k}}^{j}(\vec{r}) \tag{528}$$

To simplify the writing, we can conveniently note

$$|\Psi\rangle = \sum_{n,j,\vec{k}} a_{n,j}(\vec{k}) |\phi_{n,\vec{k}}^{j}\rangle = \sum_{n,j,\vec{k}} a_{n,j}(\vec{k}) |n,\vec{k}\rangle^{j} \tag{529}$$

Explicitly,

$$\langle \vec{r} | n,\vec{k} \rangle^{j} = e^{i\vec{k}\cdot\vec{r}} \; \phi_{n}^{j}(\vec{r}) = e^{i\vec{k}\cdot\vec{r}} \langle \vec{r} | n \rangle^{j} \tag{530}$$

if

$$\langle \vec{r} | n \rangle^{j} = \phi_{n}^{j}(\vec{r}) \tag{531}$$

B.   Zero External Magnetic Field

If $\hat{H}_0$ is the Hamiltonian operator of the system for zero magnetic induction, that is, $\hat{H}_m = \hat{0}$, assuming the functions

$$\phi_{n,\vec{k}}^j(\vec{r})$$

to be eigenfunctions of this operator for the degenerate eigenvalue $\varepsilon_n(\vec{k})$,

$$\hat{H}_0 |n,\vec{k}\rangle^j = \varepsilon_n(\vec{k}) |n,\vec{k}\rangle^j \tag{532}$$

Let us evaluate the scalar product

$$\langle n,\vec{k}|^j \hat{H}_0 \Psi\rangle = \sum_{n',j',\vec{k}'} a_{n',j'}(\vec{k}') \, \langle n,\vec{k}|^j \hat{H}_0 |n',\vec{k}'\rangle^{j'} \tag{533}$$

$$= \sum_{n',j',\vec{k}'} a_{n',j'}(\vec{k}')\varepsilon_{n'}(\vec{k}') \, \delta_{jj'} \, \delta_{nn'} \, \delta(\vec{k}' - \vec{k}) $$

hence,

$$\langle n,\vec{k}|^j \hat{H}_0 \Psi\rangle = \varepsilon_n(\vec{k}) a_{n,j}(\vec{k}) \tag{534}$$

but also, since

$$\hat{H}_0 = -\frac{\hbar^2}{2m}\nabla^2 = q\hat{V} \tag{513}$$

$$\langle n,\vec{k}|^j \hat{H}_0 \Psi\rangle = \sum_{n',j',\vec{k}'} a_{n',j'}(\vec{k}')\langle n,\vec{k}|^j -\frac{\hbar^2}{2m}\nabla^2 + q\hat{V} |n',\vec{k}'\rangle^{j'} \tag{535}$$

One can easily verify that

$$\nabla^2\left[e^{i\vec{k}'\cdot\vec{r}}\phi_{n'}^{j'}(\vec{r})\right] = e^{i\vec{k}'\cdot\vec{r}}\left[-k'^2 + 2i\vec{k}'\cdot\vec{\nabla} + \nabla^2\right]\phi_{n'}^{j'}(\vec{r}) \tag{536}$$

and since

$$\hat{\vec{p}} = -i\hbar\vec{\nabla} \tag{503}$$

this means

$$\langle {}^{j}_{n}, \vec{k} | \hat{H}_0 | {}^{j'}_{n'}, \vec{k'} \rangle = \int e^{i(\vec{k'}-\vec{k})\cdot\vec{r}} \left(\phi^{j}_{n}\right)^* \left[\frac{\hbar^2 k'^2}{2m} + \frac{\hbar}{m}\vec{k'}\cdot\vec{p} + \varepsilon_{n'}(\vec{k'})\right]\phi^{j'}_{n'} \, d^3r$$

(537)

Then the expression

$$X = \left(\phi^{j}_{n}(\vec{r})\right)^* \left[\frac{\hbar^2 k'^2}{2m} + \frac{\hbar}{m}\vec{k'}\cdot\vec{p} + \varepsilon_{n'}(\vec{k'})\right]\phi^{j'}_{n'}(\vec{r})$$

(538)

is periodic; its period is that of the graphite space lattice.
Hence, X can be developed in Fourier series in the reciprocal space

$$X = \sum_{m} C_{m}^{{}^{n',j',\vec{k'}}}{}_{{}^{n,j,\vec{k}}} \, e^{-i\vec{K}_m\cdot\vec{r}}$$

(539)

with

$$C_{m}^{{}^{n',j',\vec{k'}}}{}_{{}^{n,j,\vec{k}}} = \frac{1}{\Omega}\int_{\Omega} X e^{i\vec{K}_m\cdot\vec{r}} \, d^3r$$

(540)

Equation (537) is transformed into

$$\langle {}^{j}_{n}, \vec{k} | \hat{H}_0 | {}^{j'}_{n'}, \vec{k'} \rangle = \Omega \sum_{m} C_{m}^{{}^{n',j',\vec{k'}}}{}_{{}^{n,j,\vec{k}}} \, \delta(\vec{k'} - \vec{k} - \vec{K}_m)$$

(541)

In the first Brillouin zone $\vec{K}_m = \vec{0}$, that is, m = 0 and, by using
the property of the Dirac δ function,

$$f(x) \, \delta(x - x_0) = f(x_0) \, \delta(x - x_0)$$

(542)

$$\langle {}^{j}_{n}, \vec{k} | \hat{H}_0 | {}^{j'}_{n'}, \vec{k'} \rangle = \Omega C_{0}^{{}^{n',j',\vec{k}}}{}_{{}^{n,j,\vec{k}}} \, \delta(\vec{k'} - \vec{k})$$

(543)

It is now necessary to use (540) to calculate

$$C_{0}^{{}^{n',j',\vec{k}}}{}_{{}^{n,j,\vec{k}}} = \frac{1}{\Omega}\int_{\Omega} \left(\phi^{j}_{n}(\vec{r})\right)^* \left[\frac{\hbar^2 k'^2}{2m} + \frac{\hbar}{m}\vec{k'}\cdot\vec{p} + \varepsilon_{n'}(\vec{k'})\right]\phi^{j'}_{n'}(\vec{r}) \, d^3r$$

(544)

Three integrals must be evaluated:

$$\langle {}^{j}_{n} | \frac{\hbar^2 \hat{k}'^2}{2m} | {}^{j'}_{n'} \rangle = \frac{\hbar^2 k'^2}{2m} \, \delta_{jj'} \, \delta_{nn'}$$

(545)

$$\langle {}^{j}_{n} | \hat{\epsilon}_{n'}(\vec{k}') | {}^{j'}_{n'} \rangle = \epsilon_{n'}(\vec{k}') \, \delta_{jj'} \, \delta_{nn'} \tag{546}$$

$$\langle {}^{j}_{n} | \frac{\widehat{\cancel{h}\vec{k}' \cdot \vec{p}}}{m} | {}^{j'}_{n'} \rangle = -\frac{i\cancel{h}^2}{m} \langle {}^{j}_{n} | \sum_{\alpha} k'_{\alpha} \frac{\partial}{\partial x_{\alpha}} | {}^{j'}_{n'} \rangle \tag{547}$$

if

$$\sum_{\alpha} k_{\alpha} \frac{\partial}{\partial x_{\alpha}} = k_x \frac{\partial}{\partial x} + k_y \frac{\partial}{\partial y} + k_z \frac{\partial}{\partial z} \tag{548}$$

Let us set by definition

$$p_{\alpha}{}^{n',j'}_{} {}^{n,\,j} = \langle {}^{j}_{n} | -i\cancel{h} \frac{\partial}{\partial x_{\alpha}} | {}^{j'}_{n'} \rangle \tag{549}$$

All these results, introduced into (544), serve to write

$$C_0{}^{n',j',\vec{k}'}_{} {}^{n,\,j,\,\vec{k}} = \frac{1}{\Omega} \left[\left(\epsilon_{n'}(\vec{k}') + \frac{\cancel{h}^2 k'^2}{2m} \right) \delta_{jj'} \, \delta_{nn'} + \frac{\cancel{h}}{m} \sum_{\alpha} k'_{\alpha} p_{\alpha}{}^{n',j'}_{} {}^{n,\,j} \right] \tag{550}$$

and (543) becomes

$$\langle {}^{j}_{n}, \vec{k} | \hat{H}_0 | {}^{j'}_{n'}, \vec{k}' \rangle = \left(\epsilon_{n'}(\vec{k}') + \frac{\cancel{h}^2 k'^2}{2m} \right) \delta_{jj'} \, \delta_{nn'} \, \delta(\vec{k}' - \vec{k})$$

$$+ \frac{\cancel{h}}{m} \sum_{\alpha} k'_{\alpha} p_{\alpha}{}^{n',j'}_{} {}^{n,\,j} \, \delta(\vec{k}' - \vec{k}) \tag{551}$$

This equation facilitates the transcription of (534) and (535) into

$$\sum_{n',j',k'} \left[\left(\epsilon_{n'}(k') + \frac{\cancel{h}^2 k'^2}{2m} \right) \delta_{jj'} \, \delta_{nn'} + \frac{\cancel{h}}{m} \sum_{\alpha} k'_{\alpha} p_{\alpha}{}^{n',j'}_{} {}^{n,\,j} \right] \delta(\vec{k}' - \vec{k}) a_{n',j'}(\vec{k}')$$

$$= \epsilon_n(\vec{k}) a_{n,j}(\vec{k}) \tag{552}$$

an intermediate result, very useful for further developments, and valid for graphite in the absence of a magnetic field and impurities.

C. Analysis of Matrix Elements of the Operator $\hat{\vec{p}}$

In Eq. (552), the terms in which the matrix elements $p_{\alpha}{}^{n',j'}_{} {}^{n,\,j}$ appear represent interactions between the different bands n and n'. As a rule, the pulse $\vec{p}$ of a particle may be connected to its total mechanical energy ε by an equation of the type

$$\varepsilon = \frac{p^2}{2m} = \frac{\hbar^2 k^2}{2m} \tag{553}$$

which we can place in the form

$$p = \frac{m}{\hbar} \frac{\partial \varepsilon}{\partial k} \tag{554}$$

since

$$\vec{p} = \hbar \vec{k}$$

(Louis de Broglie correspondence equation). In (553), m is the real mass (case of a free particle) or the effective mass (general case). At the corner of the Brillouin zone, the energy is a minimum [22-24] in the case of graphite. This means that

$$p_\alpha^{\substack{n,\,j\\n',\,j'}} = 0 \tag{555}$$

at point K of this zone. Since the operator $\hat{\vec{p}}$ is hermitic, such that

$$\hat{\vec{p}} = \hat{\vec{p}}^+ \tag{556}$$

the matrix elements of $\hat{\vec{p}}$ obey the following relation:

$$p_\alpha^{\substack{n,\,j\\n',\,j'}} = \left(\widetilde{p_\alpha^{\substack{n,\,j\\n',\,j'}}} \right)^* = \left(p_\alpha^{\substack{n',\,j'\\n,\,j}} \right)^* \tag{557}$$

Note that the operator $\hat{\vec{p}}$ is odd:

$$\hat{p}_\alpha \equiv -i\hbar \frac{\partial}{\partial x_\alpha}$$

becomes $-\hat{p}_\alpha$ if we change x_α in $-x_\alpha$. We can obtain an important property from this. If we note $\hat{P}$ the parity operator such that

$$\hat{P} |\vec{r}\rangle = |-\vec{r}\rangle \tag{558}$$

and

$$\langle \vec{r} | \hat{P}\Psi \rangle = \Psi(-\vec{r}) \tag{559}$$

we easily have

$$\hat{P}^2 = \hat{1} \qquad \text{that is,} \qquad \hat{P} = \hat{P}^+ = \hat{P}^{-1} \tag{560}$$

The parity operator is unitary.

An operator $\hat{O}$ subjected to $\hat{P}$ is transformed into $\hat{O}'$ such that

$$\hat{O}' = \hat{P}^{-1} \hat{O} \hat{P} = \hat{P} \hat{O} \hat{P} \tag{561}$$

If $\hat{O}' = +\hat{O}$, the operator $\hat{O}$ is called even.

If $\hat{O}' = -\hat{O}$, the operator $\hat{O}$ is called odd.

The remark made above about $\hat{\vec{p}}$ implies that

$$\hat{P} \hat{\vec{p}} \hat{P} = -\hat{\vec{p}} \tag{562}$$

Let us denote $\Phi(\vec{r})$ and $\Psi(\vec{r})$ two functions of the same parity. Hence they obey

$$\Phi'(\vec{r}) = \hat{P}\Phi(\vec{r}) = \Phi(\pm\vec{r})$$
$$\Psi'(\vec{r}) = \hat{P}\Psi(\vec{r}) = \Psi(\pm r) \tag{563}$$

and hence,

$$<\Phi|\hat{\vec{p}}\Psi> = -<\Phi|\hat{P}\hat{\vec{p}}\hat{P}\Psi> = -\Phi'|\hat{\vec{p}}\Psi'> \tag{564}$$

since

$$<\Psi|\hat{P} = <\hat{P}^+\Psi| = <\hat{P}\Psi| = <\Psi'| \tag{565}$$

In conclusion, from (564), the matrix elements of an odd operator are zero between two vectors of the same parity. In the base of vectors

$$\left\{ |_{n,\vec{k}}^{\,j}> \right\}$$

the functions

$$\phi_{n,k}^{j}(\vec{r}) \qquad \text{and} \qquad \phi_{n,k}^{j'}(\vec{r})$$

belong to the same irreducible representation [11-14]. They have the same parity and hence

$$p_\alpha^{n,j'}{}^{n,j} = 0 \qquad \text{for all values of } j \text{ and } j' \tag{566}$$

In representation $\vec{r}$, that is, in the base of the Dirac δ functions centered on different points in space, eigenfunctions of the operator $\hat{\vec{r}}$, the pulse operator has the functional expression [29]

$$\hat{\vec{p}} = -i\hbar\vec{\nabla} \tag{503}$$

In representation $\vec{p}$ [51,52], any wave function is given by a development of the eigenfunctions of the operator $\hat{\vec{p}}$, that is, on the plane wave functions

$$\Psi(\vec{r}) = \int a(\vec{p})\Psi_{\vec{p}}(\vec{r}) \, d^3p \tag{567}$$

where

$$\Psi_{\vec{p}}(\vec{r}) = \frac{1}{(2\pi\hbar)^{3/2}} e^{\frac{i}{\hbar}\vec{p}\cdot\vec{r}} \tag{568}$$

In the base of $\{|\Psi_{\vec{p}}(\vec{r})>\}$; a reasoning similar to that which led to Eq. (503) give us the functional expression of the operator $\hat{\vec{r}}$:

$$\hat{\vec{r}} = i\hbar \frac{\partial}{\partial\vec{p}} \tag{569}$$

For further details, a simple demonstration is provided in Refs. 17, 18, 51, and 52.

D. Equivalence Formulas of Luttinger and Kohn

1. *Eigenvalue Equation*

In the presence of a constant magnetic field, the Hamiltonian operator $\hat{H}$ is in the form

$$\hat{H} = \hat{H}_0 + \hat{H}_m \tag{570}$$

where the term $\hat{H}_m$ corresponds to Eq. (517).

If Ψ is the wave function of the system, it is necessary to solve the equation

$$\hat{H}|\Psi> = \varepsilon|\Psi> \tag{571}$$

Hence by multiplying at the left by the bra $\langle n,\vec{k}|^{j}$ and using (552),

$$\sum_{n',j',\vec{k}'}\left[\left(\varepsilon_{n'}(\vec{k}') + \frac{\hbar^2 k'^2}{2m}\right)\delta_{jj'}\,\delta_{nn'} + \frac{\hbar}{m}\sum_{\alpha}k'_{\alpha}p_{\alpha}^{n',j'}\Big|^{n,j}\right]\delta(\vec{k}' - \vec{k})a_{n',j'}(\vec{k}')$$

$$+ \,^{j}\langle n,\vec{k}|\hat{H}_m\Psi\rangle = \varepsilon a_{n,j}(\vec{k}) \tag{572}$$

Let us evaluate the scalar product

$$^{j}\langle n,\vec{k}|\hat{H}_m|\Psi\rangle = \,^{j}\langle n,\vec{k}|\frac{q^2}{2m}B^2\hat{y}^2 + \frac{q}{m}B\hat{y}\hat{p}_x|\Psi\rangle \tag{573}$$

$$^{j}\langle n,\vec{k}|\frac{q^2}{2m}B^2\hat{y}^2|\Psi\rangle = \frac{q^2}{2m}B^2\sum_{n',j',\vec{k}'}\,^{j}\langle n,\vec{k}|\hat{y}^2|n',\vec{k}'\rangle^{j'}a_{n',j'}(\vec{k}') \tag{574}$$

since

$$^{j}\langle n,\vec{k}|\hat{y}^2|n',\vec{k}'\rangle^{j'} = \int e^{i(\vec{k}'-\vec{k})\cdot\vec{r}}\,y^2\left(\phi_n^j\right)^*\phi_{n'}^{j'}\,d^3r \tag{575}$$

and

$$\frac{\partial^2}{\partial k'^2_y}\int e^{i(\vec{k}'-\vec{k})\cdot\vec{r}}\left(\phi_n^j\right)^*\phi_{n'}^{j'}\,d^3r = -\int y^2 e^{i(\vec{k}'-\vec{k})\cdot\vec{r}}\left(\phi_n^j\right)^*\phi_{n'}^{j'}\,d^3r \tag{576}$$

$$^{j}\langle n,\vec{k}|\hat{y}^2|n',\vec{k}'\rangle^{j'} = \frac{\partial^2}{\partial k'^2_y}\int e^{i(\vec{k}'-\vec{k})\cdot\vec{r}}\,Y\,d^3r \tag{577}$$

By a similar reasoning to that which led us from (537) to (551), if

$$Y = \left(\phi_n^j(\vec{r})\right)^*\phi_{n'}^{j'}(\vec{r}) \tag{578}$$

it is easy to demonstrate (527) and the identity

$$^{j}\langle n,\vec{k}|\frac{q^2}{2m}B^2\hat{y}^2\Psi\rangle = -\frac{q^2}{2m}B^2\sum_{n',j',\vec{k}'}\left[\frac{\partial^2}{\partial k'^2_y}\delta(\vec{k}' - \vec{k})\right]\delta_{jj'}\,\delta_{nn'}\,a_{n',j'}(\vec{k}') \tag{579}$$

$$^{j}\langle n,\vec{k}|\frac{q}{m}B\hat{y}\hat{p}_x\Psi\rangle = -\frac{i\hbar}{m}qB\sum_{n',j',\vec{k}'}\,^{j}\langle n,\vec{k}|y\frac{\partial}{\partial x}|n',\vec{k}'\rangle^{j'}a_{n',j'}(\vec{k}') \tag{580}$$

since

$$\frac{\partial}{\partial x}\left(e^{i\vec{k}'\cdot\vec{r}}\phi_{n'}^{j'}\right) = e^{i\vec{k}'\cdot\vec{r}}\left[ik_x' + \frac{\partial}{\partial x}\right]\phi_{n'}^{j'} \tag{581}$$

and

$$\frac{\partial}{\partial k_y'}\left[k_x' e^{i(\vec{k}'-\vec{k})\cdot\vec{r}}\left(\phi_n^j\right)^* \phi_{n'}^{j'}\right] = iy k_x'\left(\phi_{n,\vec{k}}^j\right)^* \phi_{n',\vec{k}'}^{j'} \tag{582}$$

and also

$$\frac{\partial}{\partial k_y'}\left[e^{i(\vec{k}'-\vec{k})\cdot\vec{r}}\left(\phi_n^j\right)^* \frac{\partial}{\partial x}\phi_{n'}^{j'}\right] = iy\left(\phi_{n,\vec{k}}^j\right)^* \frac{\partial}{\partial x}\phi_{n',\vec{k}'}^{j'} \tag{583}$$

leading to

$$\langle{}_{n,\vec{k}}^{j}|q\frac{B}{m}\hat{y}\hat{p}_x\Psi\rangle = -\frac{i\hbar}{m}qB\sum_{n',j',\vec{k}'}\left\{k_x'\left[\frac{\partial}{\partial k_y'}\delta(\vec{k}' - \vec{k})\right]\delta_{jj'}\,\delta_{nn'}\right.$$

$$\left. - i\frac{\partial}{\partial k_y'}\langle{}_{n,\vec{k}}^{j}|\frac{\partial}{\partial x}|{}_{n',\vec{k}'}^{j'}\rangle\right\}a_{n',j'}(\vec{k}') \tag{584}$$

The integral

$$\langle{}_{n,\vec{k}}^{j}|\frac{\partial}{\partial x}|{}_{n',\vec{k}'}^{j'}\rangle$$

is calculated by setting

$$Z = \left(\phi_n^j\right)^* \frac{\partial}{\partial x}\phi_{n'}^{j'} \tag{585}$$

and by applying the method described from (537) to (551):

$$\int e^{i(\vec{k}'-\vec{k})\cdot\vec{r}} Z\, d^3r = \frac{i}{\hbar}p_x{}^{n',j'}_{}{}^{n,j}\,\delta(\vec{k}' - \vec{k}) \tag{586}$$

[see definition (549)]. With all calculations made,

$$\langle{}_{n,\vec{k}}^{j}|q\frac{B}{m}\hat{y}\hat{p}_x\Psi\rangle = -\frac{i\hbar}{m}qB\sum_{n',j',\vec{k}'}\left[\frac{\partial}{\partial k_y'}\delta(\vec{k}' - \vec{k})\right]\left[k_x'\,\delta_{jj'}\,\delta_{nn'}\right.$$

$$\left. + \frac{p_x{}^{n',j'}_{}{}^{n,j}}{\hbar}\right]a_{n',j'}(\vec{k}') \tag{587}$$

The intermediate results (587) and (579) are introduced into (572) to give

$$
\sum_{n',j',\vec{k}'} \left\{ \delta_{jj'} \; \delta_{nn'} \left[\left(\varepsilon_{n'}(\vec{k}') + \frac{\hbar^2 k'^2}{2m} \right) \delta(\vec{k}' - \vec{k}) - \frac{i\hbar}{m} qBk'_x \frac{\partial}{\partial k'_y} \delta(\vec{k}' - \vec{k}) \right. \right.
$$

$$
\left. - \frac{q^2 B^2}{2m} \frac{\partial^2}{\partial k'^2_y} \delta(\vec{k}' - \vec{k}) \right] + \frac{\hbar}{m} \sum_{\alpha} k'_\alpha p_{\alpha}{}^{n',j'}_{n,j} \delta(\vec{k}' - \vec{k})
$$

$$
\left. - \frac{iB}{m} q p_x{}^{n',j'}_{n,j} \frac{\partial}{\partial k'_y} \delta(\vec{k}' - \vec{k}) \right\} a_{n',j'}(\vec{k}') = \varepsilon a_{n,j}(\vec{k}) \tag{588}
$$

If $B = 0$, this obviously leads to (552). The first terms of (588) are multiplied by $\delta_{jj'}$, $\delta_{nn'}$. They correspond to the n band. The last two, which involve matrix elements of the operator $\hat{p}$, are interaction terms between the energy bands for the same value of $\vec{k}$.

2. Base Change

Let us now take a closer look at (588). This identity may be considered an equation for finding eigenvalues, coefficients $a_{n,j}(\vec{k})$ playing the role of eigenvalues. Symbolically we can write

$$
\hat{H}|a\rangle = \varepsilon |a\rangle \tag{589}
$$

To solve (589) it would be convenient, at the first order in $\vec{k}$, to suppress the coupling terms between bands, that is, the expressions containing matrix elements of the operators $\hat{p}$ and $\hat{p}_x$.

Let us therefore try to determine a unitary transformation $\hat{T}$ that, applied to the vectors of the base previously used, suppresses the transitions between bands at the first order in $\vec{k}$.

By definition, a unitary transformation [1,2]

$$
\hat{T}^+ = \hat{T}^{-1} = \hat{T} \tag{590}
$$

if $\hat{T}$ is hermitic. The $a_{n,j}(\vec{k})$ are transformed into $A_{n,j}(\vec{k})$ according to the equation

$$
|A\rangle = \hat{T}|a\rangle \qquad \text{or} \qquad |a\rangle = \hat{T}|A\rangle \tag{591}
$$

and, by taking the operator T in the form of a function of operator $\hat{S}$

$$\hat{T} = e^{\hat{S}} = \hat{1} + \hat{S} + \frac{\hat{S}^2}{2!} + \cdots + \frac{\hat{S}^n}{n!} + \cdots \tag{592}$$

Equation (589) becomes

$$\hat{H}\left|e^{\hat{S}}A\right> = \varepsilon\left|e^{\hat{S}}A\right> \tag{593}$$

and, by multiplying on the left by $e^{-\hat{S}}$,

$$e^{-\hat{S}}\hat{H}e^{\hat{S}}|A> = \varepsilon|A> \tag{594}$$

Let us now set

$$\hat{H}' = e^{-\hat{S}}\hat{H}e^{\hat{S}} \tag{595}$$

Since by definition of the commutator,

$$[\hat{H}, \hat{S}] = \hat{H}\hat{S} - \hat{S}\hat{H}$$

and it is easily verified that

$$[[\hat{H}, \hat{S}], \hat{S}] = \hat{H}\hat{S}^2 - 2\hat{S}\hat{H}\hat{S} + \hat{S}^2\hat{H} \tag{596}$$

We have

$$\hat{H}' = \hat{H} + [\hat{H}, \hat{S}] + \frac{1}{2}[[\hat{H}, \hat{S}], \hat{S}] + \cdots \tag{597}$$

Before making the $\hat{T}$ transformation, we can work in the base of $\{|_{n,\vec{k}}^{j}>\}$ defined by (530). Any function $\Psi(\vec{r})$ may be developed on this base [see Eq. (529)]:

$$|\Psi> = \sum_{n',j',\vec{k}'} a_{n',j'}(\vec{k}')|_{n',\vec{k}'}^{j'}>$$

with

$$a_{n,j}(\vec{k}) = <_{n,\vec{k}}^{j}|\Psi> \tag{598}$$

In the new base, obtained from $|_{n,\vec{k}}^{j}>$ by the application of the canonical transformation $\hat{T}$, the unitary vectors are written $\hat{T}|_{n,\vec{k}}^{j}>$, and the equivalent of (529) is

$$|\Psi> = \sum_{n',j',\vec{k}'} A_{n',j'}(\vec{k}')\hat{T}|_{n',\vec{k}'}^{j'}>$$ (599)

From (598) and (599),

$$a_{n,j}(\vec{k}) = \sum_{n',j',\vec{k}'} <_{n,\vec{k}}^{j}|\hat{T}|_{n',\vec{k}'}^{j'}>A_{n',j'}(\vec{k}')$$

or

$$a_{n,j}(\vec{k}) = \sum_{n',j'} <_{n}^{j}|\hat{T}|_{n'}^{j'}>A_{n',j'}(\vec{k})$$ (600)

Hence we now attempt to find the transformation $\hat{T}$ (in fact, the matrix elements of $\hat{S}$) enabling us to write (589) in the form

$$\hat{H}'|A> = \varepsilon|A>$$ (601)

an equation in which, by a judicious choice of $\hat{T}$, the transitions between bands have been eliminated at the first order in $\vec{k}$.

Let

$$\hat{H} = \hat{H}_1 + \hat{H}_2 + \hat{H}_3 + \hat{H}_4$$ (602)

with $\hat{H}_1$, $\hat{H}_2$, $\hat{H}_3$, and $\hat{H}_4$ operators such that [see (551) and (588)]

$$<_{n,\vec{k}}^{j}|\hat{H}_1|_{n',\vec{k}'}^{j'}> = \left[\varepsilon_{n'}(\vec{k}') + \frac{\hbar^2 k'^2}{2m}\right]\delta_{jj'}\,\delta_{nn'}\,\delta(\vec{k}' - \vec{k})$$ (603)

$$<_{n,\vec{k}}^{j}|\hat{H}_2|_{n',\vec{k}'}^{j'}> = \frac{\hbar}{m}\sum_{\alpha} k'_\alpha p_\alpha^{n',j'\,n,j}\,\delta(\vec{k}' - \vec{k}) - iq\frac{B}{m}p_x^{n',j'\,n,j}\frac{\partial}{\partial k'_y}\delta(\vec{k}' - \vec{k})$$ (604)

$$<_{n,\vec{k}}^{j}|\hat{H}_3|_{n',\vec{k}'}^{j'}> = -\frac{i\hbar}{m}qBk'_x\,\delta_{jj'}\,\delta_{nn'}\frac{\partial}{\partial k'_y}\delta(\vec{k}' - \vec{k})$$ (605)

$$<_{n,\vec{k}}^{j}|\hat{H}_4|_{n',\vec{k}'}^{j'}> = -q^2\frac{B^2}{2m}\,\delta_{jj'}\,\delta_{nn'}\frac{\partial^2}{\partial k'^2_y}\delta(\vec{k}' - \vec{k})$$ (606)

The transform of the Hamiltonian operator $\hat{H}$ by the base change $\hat{T}$ (or $\hat{S}$) is, according to (597),

$$\hat{H}' = \sum_{\ell=1}^{4} \{\hat{H}_\ell + [\hat{H}_\ell,\ \hat{S}] + \frac{1}{2}[[\hat{H}_\ell,\ \hat{S}],\ \hat{S}] + \cdots\} \tag{607}$$

Let us restate the problem. We are attempting to find the unitary transformation $\hat{T}$ (i.e., the matrix elements of the operator $\hat{S}$) enabling us to eliminate the interaction terms between bands in Eq. (588). To suppress them at the first order in $\vec{k}$, Luttinger and Kohn impose the condition

$$\hat{H}_2 + [\hat{H}_1,\ \hat{S}] = \hat{0} \tag{608}$$

this modifies the writing of (607) to give

$$\hat{H}' = \hat{H}_1 + \hat{H}_3 + \hat{H}_4 + \frac{1}{2}[\hat{H}_2,\ \hat{S} + [\hat{H}_3,\ \hat{S}] + [\hat{H}_4,\ \hat{S}] + \frac{1}{2}[[\hat{H}_2,\ \hat{S}],\ \hat{S}]$$

$$+ \frac{1}{2}[[\hat{H}_3,\ \hat{S}],\ \hat{S}] + \frac{1}{2}[[\hat{H}_4,\ \hat{S}],\ \hat{S}] + \cdots \tag{609}$$

3. Matrix Elements of $\hat{S}$

Condition (608) provides the scalar product

$$\langle {}^{j}_{n,\vec{k}}|\hat{H}_2 + \hat{H}_1\hat{S} - \hat{S}\hat{H}_1|{}^{j'}_{n',\vec{k}'}\rangle = 0 \tag{610}$$

The evaluations of the different terms appearing in (610) are simplified by applying the so-called base closure equation of the base $\{|{}^{j}_{n,\vec{k}}\rangle\}$

$$\sum_{n'',j'',\vec{k}''} |{}^{j''}_{n'',\vec{k}''}\rangle\langle {}^{j''}_{n'',\vec{k}''}| = \hat{1} \tag{611}$$

$$\langle {}^{j}_{n,\vec{k}}|\hat{H}_1\hat{S}|{}^{j'}_{n',\vec{k}'}\rangle = \sum_{n'',j'',\vec{k}''} \left[\varepsilon_{n''}(k'') + \frac{\hbar^2 k''^2}{2m}\right] \delta_{jj''}\ \delta_{nn''}\ \delta(\vec{k}'' - \vec{k})$$

$$\cdot\ \langle {}^{j''}_{n'',\vec{k}''}|\hat{S}|{}^{j'}_{n',\vec{k}'}\rangle = \left[\varepsilon_n(\vec{k}) + \frac{\hbar^2 k^2}{2m}\right]\langle {}^{j}_{n,\vec{k}}|\hat{S}|{}^{j'}_{n',\vec{k}'}\rangle \tag{612}$$

By a similar calculation,

$$\langle {}^{j}_{n,\vec{k}}|\hat{S}\hat{H}_1|{}^{j'}_{n',\vec{k}'}\rangle = \left[\varepsilon_{n'}(\vec{k}') + \frac{\hbar^2 k'^2}{2m}\right]\langle {}^{j}_{n,\vec{k}}|\hat{S}|{}^{j'}_{n',\vec{k}'}\rangle \tag{613}$$

The use of Eqs. (604), (612), and (613) in (610) yields the matrix elements of the canonical transformation. Let

$$\hbar\omega_{nn'} = \varepsilon_n(\vec{k}) - \varepsilon_{n'}(\vec{k}') + \frac{\hbar^2}{2m}(k^2 - k'^2) \tag{614}$$

and hence,

$$<^j_{n,\vec{k}}|\hat{S}|^{j'}_{n',\vec{k}'}> = -\frac{<^j_{n,\vec{k}}|\hat{H}_2|^{j'}_{n',\vec{k}'}>}{\hbar\omega_{nn'}} \tag{615}$$

Thus finally,

$$<^j_{n,\vec{k}}|\hat{S}|^{j'}_{n',\vec{k}'}> = -\frac{1}{m\omega_{nn'}}\left[\sum_\alpha k'_\alpha p^{n,j}_\alpha{}^{n',j'}\,\delta(\vec{k}' - \vec{k})\right.$$

$$\left. - iq\frac{B}{\hbar}p^{n,j}_x{}^{n',j'}\frac{\partial}{\partial k'_y}\,\delta(\vec{k}' - \vec{k})\right] \tag{616}$$

At this stage of the calculation, it is advisable to review the properties of the matrix elements of the pulse operator $\hat{p}$: if $n = n'$ and $j = j'$ or $j \neq j'$ simultaneously, then,

$$<^j_{n,\vec{k}}|\hat{S}|^{j'}_{n,\vec{k}'}> = 0 \tag{617}$$

4. Effective Mass

Let us reconsider Eq. (571) and multiply at the left by the bra $<^j_{n,\vec{k}}|$:

$$\sum_{n',j',\vec{k}'} a_{n',j'}(\vec{k}')<^j_{n,\vec{k}}|\hat{H}|^{j'}_{n',\vec{k}'}> = \varepsilon a_{n,j}(\vec{k}) \tag{618}$$

By the canonical transformation $\hat{T}$ this equation becomes [see (601)]

$$\sum_{n',j',\vec{k}'} A_{n',j'}(\vec{k}')<^j_{n,\vec{k}}|\hat{H}'|^{j'}_{n',\vec{k}'}> = \varepsilon A_{n,j}(\vec{k}) \tag{619}$$

the expression of the operators $\hat{H}$ being that of (609). The different terms making up the scalar product $<^j_{n,\vec{k}}|\hat{H}'|^{j'}_{n',\vec{k}'}>$ are supplied by

(603) to (606) and (616). But if we are only interested in the case n' = n, that is, in the interactions in a single band,

$$\langle {}^j_{n,\vec{k}}|\,[\hat{H}_3,\ \hat{S}]\,|{}^{j'}_{n,\vec{k}'}\rangle = \sum_{n'',j'',\vec{k}''}\left\{\langle {}^j_{n,\vec{k}}|\hat{H}_3|{}^{j''}_{n'',\vec{k}''}\rangle\langle {}^{j''}_{n'',\vec{k}''}|\hat{S}|{}^{j'}_{n,\vec{k}'}\rangle\right.$$

$$\left. - \langle {}^j_{n,\vec{k}}|\hat{S}|{}^{j''}_{n'',\vec{k}''}\rangle\langle {}^{j''}_{n'',\vec{k}''}|\hat{H}_3|{}^{j'}_{n,\vec{k}'}\rangle\right\} \qquad (620)$$

According to (605), these terms are proportional to $\delta_{jj''}\delta_{nn''}$ or $\delta_{j'j''}\delta_{nn''}$. Hence if n = n'', we know that the matrix elements of the operator $\hat{S}$ are zero [see (617)], and

$$\langle {}^j_{n,\vec{k}}|\,[\hat{H}_3,\ \hat{S}]\,|{}^{j'}_{n,\vec{k}'}\rangle = 0 \qquad (621)$$

The same applies to $\langle {}^j_{n,\vec{k}}|\,[\hat{H}_4,\ \hat{S}]\,|{}^{j'}_{n,\vec{k}'}\rangle$ since (606) clearly indicates that the matrix elements of H_4 are proportional to $\delta_{nn'}$. This means that in the expression of $\hat{H}'$ [Eq. (609)], the only second-order term in $\hat{S}$ is the commutator $\frac{1}{2}[\hat{H}_2,\ \hat{S}]$ and (619) written by ignoring all the other terms, which are at least of the third order in S:

$$\sum_{j',k'} A_{n,j'}(\vec{k}')\langle {}^j_{n,\vec{k}}|\hat{H}_1 + \hat{H}_3 + \hat{H}_4 + \tfrac{1}{2}[\hat{H}_2,\ \hat{S}]\,|{}^{j'}_{n,\vec{k}'}\rangle = \varepsilon A_{n,j}(\vec{k})$$
$$(622)$$

Using (604), (615), and the closure equation (611), we have

$$\tfrac{1}{2}\langle {}^j_{n,\vec{k}}|\,[\hat{H}_2,\ \hat{S}]\,|{}^{j'}_{n,\vec{k}'}\rangle = \sum_{\substack{n'',j'',\vec{k}'' \\ \alpha,\beta}} \frac{1}{\hbar m^2\omega_{nn''}}\left\{\hbar^2 k''_\alpha k'_\beta p^{n,\;j}_{\;\;n'',j''}p^{n'',j''}_{\;\;n',j'}\,\delta(\vec{k}''-\vec{k})\,\delta(\vec{k}'-\vec{k}'')\right.$$

$$- q^2 B^2 p^{n,\;j}_{\;\;n'',j''}p^{n'',j''}_{\;\;n',j'}\,\frac{\partial}{\partial k'_y}\delta(\vec{k}'-\vec{k}'')\,\frac{\partial}{\partial k''_y}\delta(\vec{k}''-\vec{k}) - iqB\hbar\left[k''_\alpha p^{n,\;j}_{\;\;n'',j''}p^{n'',j''}_{\;\;n',j'}\right.$$

$$\left.\left. \cdot\,\delta(\vec{k}''-\vec{k})\,\frac{\partial}{\partial k'_y}\delta(\vec{k}'-\vec{k}'') + k'_\alpha p^{n'',j''}_{\;\;n',j'}p^{n,\;j}_{\;\;n'',j''}\,\delta(\vec{k}'-\vec{k}'')\,\frac{\partial}{\partial k''_y}\delta(\vec{k}''-k)\right]\right\}$$
$$(623)$$

For graphite, the energy is a minimum at the corner K of the Brillouin zone. In the neighborhood of this point and in the absence of a magnetic field, since

$$[\hat{H}']_{B=0} = \hat{H}_1 + \frac{1}{2}\left[\hat{H}_2, \ \hat{S}\right]_{B=0} \tag{624}$$

and by taking the origin of the energies at this point, we can write [see Eqs. (622) through (624)]

$$\varepsilon_n(\vec{k}) = \frac{\hbar^2 k^2}{2m} + \sum_{\substack{n'',j'' \\ \alpha,\beta \\ n''\neq n}} \frac{\hbar}{m^2 \omega_{nn''}} k_\alpha k_\beta p_\alpha^{n,j \ n'',j''} p_\beta^{n',j'} \tag{625}$$

an equation valid in the neighborhood of the corner of the Brillouin zone.

a. *Dirac's δ function* [17,18]. One of its representations is

$$\delta(k_x) = \frac{1}{2\pi} \int_{-\infty}^{+\infty} e^{ik_x x} \ dx \tag{626}$$

This is such that

$$\int_{-\infty}^{+\infty} \delta(k_x) \ dk_x = 1 \tag{627}$$

and that

$$f(k_0) = \int_{-\infty}^{+\infty} f(k_x) \ \delta(k_x - k_0) \ dk_x \tag{628}$$

namely,

$$f(k_x) \ \delta(k_x - k_0) = f(k_0) \ \delta(k_x - k_0) \tag{629}$$

The parity of the Dirac function can also be inferred from the definition (626)

$$\delta(-k_x) = \delta(k_x) \tag{630}$$

If $\vec{k}$ and $\vec{k}'$ are two vectors with components k_x, k_y, k_z (k_x', k_y', and k_z', respectively) in a system of orthonormed axes, it is easily established from (626) that

$$\delta(k_x' - k_x) \ \delta(k_y' - k_y) \ \delta(k_z' - k_z) = \delta(\vec{k}' - \vec{k}) \tag{631}$$

Assuming

$$\delta(\vec{k}' - \vec{k}) = \frac{1}{(2\pi)^3} \ \int_{-\infty}^{+\infty} e^{i(\vec{k}'-\vec{k})\cdot\vec{r}} \ d^3r \tag{632}$$

Let us evaluate the integral

$$I = \int d^3k'' \; \delta(\vec{k} - \vec{k}'') \; \delta(\vec{k}' - \vec{k}'')$$

Equation (629), with the equivalent of (627) in three dimensions, provides

$$I = \int d^3k'' \; \delta(\vec{k} - \vec{k}') \; \delta(\vec{k}' - \vec{k}'') = \delta(\vec{k} - \vec{k}') \tag{633}$$

Let us return to Eq. (629) and assume

$$f(k_x) = k_x \qquad \text{and} \qquad k_0 = 0$$

It is immediately written

$$k_x \; \delta(\vec{k}) = 0 \tag{634}$$

Thus, by changing the writing of the variable,

$$(k_\alpha - k_\alpha') \; \delta(\vec{k} - \vec{k}') = 0 \qquad \text{if} \qquad \alpha = x, y, z \tag{635}$$

By deriving (635) in relation to k_β', we have

$$(k_\alpha - k_\alpha') \; \frac{\partial}{\partial k_\beta'} \; \delta(\vec{k} - \vec{k}') = \delta_{\alpha\beta} \; \delta(\vec{k} - \vec{k}') \tag{636}$$

Explicitly,

$$(k_x - k_x') \; \frac{\partial}{\partial k_y'} \; \delta(\vec{k} - \vec{k}') = 0 \tag{637}$$

$$(k_y - k_y') \; \frac{\partial}{\partial k_z'} \; \delta(\vec{k} - \vec{k}') = 0 \tag{638}$$

but

$$(k_x - k_x') \; \frac{\partial}{\partial k_x'} \; \delta(\vec{k} - \vec{k}') = \delta(\vec{k} - \vec{k}') \tag{639}$$

and the analogous equations obtained by circular permutation.

The representation (626) of the δ function serves to obtain the identities

$$\frac{\partial}{\partial k_y''} \; \delta(\vec{k}'' - \vec{k}) = iy \; \delta(\vec{k}'' - \vec{k}) \tag{640}$$

$$\frac{\partial^2}{\partial k_y''^2} \; \delta(\vec{k}'' - \vec{k}) = -y^2 \; \delta(\vec{k}'' - \vec{k}) \tag{641}$$

Hence by combining them,

$$\frac{\partial}{\partial k''_y} \delta(\vec{k}'' - \vec{k}) \frac{\partial}{\partial k'_y} \delta(\vec{k}' - \vec{k}'') = \delta(\vec{k}' - \vec{k}) \frac{\partial^2}{\partial k'^2_y} \delta(\vec{k}' - \vec{k}) \qquad (642)$$

b. *Effective mass tensor* [5-8]. For a free particle, the total mechanical energy ε is purely kinetic. It is expressed by

$$\varepsilon = \frac{\hbar^2 k^2}{2m} \qquad (643)$$

which serves to obtain

$$\frac{1}{m} = \frac{1}{\hbar^2} \frac{\partial^2 \varepsilon}{\partial k^2} \qquad (644)$$

By generalizing (644) to nonisolated particles, the effective mass tensor m' is defined by

$$\frac{1}{m'} = \frac{1}{\hbar^2} \frac{\partial^2}{\partial k_\alpha \partial k_\beta} \qquad (645)$$

One can easily write (625) in the form (645). This yields to the "f sum rule":

$$\frac{2}{m} \sum_{\substack{n'',j'' \\ \alpha,\beta \\ n'' \neq n}} \frac{p_\alpha^{n,j} \, {}^{n'',j''} \, p_\beta^{n'',j''} \, {}^{n',j'}}{\hbar \omega_{n''n}} = \sum_\alpha \delta_{\alpha\beta} - \frac{m}{\hbar^2} \frac{\partial^2 \varepsilon_n}{\partial k_\alpha \partial k_\beta} \qquad (646)$$

To evaluate the scalar product $\langle \begin{smallmatrix} j \\ n,\vec{k} \end{smallmatrix} | \hat{H}' | \begin{smallmatrix} j' \\ n,\vec{k}' \end{smallmatrix} \rangle$, which appears in the left-hand member of Eq. (622), we use (603), (605), (606), and (623) reformulated by using (646) and the different properties of the Dirac δ function [Eqs. (626) through (642)]

$$\langle \begin{smallmatrix} j \\ n,\vec{k} \end{smallmatrix} | \hat{H}' | \begin{smallmatrix} j' \\ n,\vec{k}' \end{smallmatrix} \rangle = \frac{1}{2} \sum_{\alpha,\beta} k_\alpha k_\beta \frac{\partial^2 \varepsilon_n}{\partial k_\alpha \partial k_\beta} \delta(\vec{k}' - \vec{k}) - iq \frac{B\hbar}{2m} \frac{\partial}{\partial k'_y} \delta(\vec{k}' - \vec{k})$$

$$\cdot \left[k_x - k'_x + \frac{m}{\hbar^2} \sum_\alpha (k_\alpha + k'_\alpha) \frac{\partial^2 \varepsilon_n}{\partial k_\alpha \partial k_x} \right] - q^2 \frac{B^2}{2\hbar^2} \frac{\partial^2}{\partial k'^2_y} \delta(\vec{k}' - \vec{k}) \frac{\partial^2 \varepsilon_n}{\partial k^2_x}$$

$$(647)$$

Since

$$(k_x - k_x') \frac{\partial}{\partial k_y'} \delta(\vec{k}' - \vec{k}) = 0 \qquad (636)$$

and

$$\sum_\alpha (k_\alpha + k_\alpha') \frac{\partial}{\partial k_y'} \delta(\vec{k}' - \vec{k}) = \sum_\alpha \left[(k_\alpha' - k_\alpha) + 2k_\alpha \right] \frac{\partial}{\partial k_y'} \delta(\vec{k}' - \vec{k})$$

$$= -\delta_{\alpha y} \delta(\vec{k}' - \vec{k}) + 2 \sum_\alpha k_\alpha \frac{\partial}{\partial k_y'} \delta(\vec{k}' - \vec{k})$$

$$(648)$$

we obtain

$$\langle {}^j_{n,\vec{k}} | \hat{H}_1 | {}^{j'}_{n,\vec{k}'} \rangle = \frac{1}{2} \sum_{\alpha,\beta} k_\alpha k_\beta \frac{\partial^2 \varepsilon_n}{\partial k_\alpha \partial k_\beta} \delta(\vec{k}' - \vec{k}) + iq \frac{B}{2\hbar} \frac{\partial^2 \varepsilon_n}{\partial k_x \partial k_y} \delta(\vec{k}' - \vec{k})$$

$$- iq \frac{B}{\hbar} \sum_\alpha k_\alpha \frac{\partial^2 \varepsilon_n}{\partial k_\alpha \partial k_x} \frac{\partial}{\partial k_y'} \delta(\vec{k}' - \vec{k}) - q^2 \frac{B^2}{2\hbar^2} \frac{\partial^2 \varepsilon_n}{\partial k_x^2} \frac{\partial^2}{\partial k_y^2} \delta(\vec{k}' - \vec{k})$$

$$(649)$$

This is equivalent to writing (626) in the form

$$\sum_{j'} \left\{ \left[\frac{1}{2} \sum_{\alpha,\beta} k_\alpha k_\beta \frac{\partial^2 \varepsilon_n}{\partial k_\alpha \partial k_\beta} + iq \frac{B}{2\hbar} \frac{\partial^2 \varepsilon_n}{\partial k_y \partial k_x} \right] A_{n,j'}(\vec{k}) \right.$$

$$- iq \frac{B}{\hbar} \sum_\alpha k_\alpha \frac{\partial^2 \varepsilon_n}{\partial k_\alpha \partial k_x} \int \frac{\partial}{\partial k_y'} \delta(\vec{k}' - \vec{k}) A_{n,j'}(\vec{k}') \, d^3 k'$$

$$\left. - \frac{q^2}{2\hbar^2} \frac{\partial^2 \varepsilon_n}{\partial k_x^2} \int \frac{\partial^2}{\partial k_y'^2} \delta(\vec{k}' - \vec{k}) A_{n,j'}(\vec{k}') \, d^3 k' \right\} = \varepsilon A_{n,j}(\vec{k}) \qquad (650)$$

Part integrations allow the substitution of the integrals by derivatives in relation to k_y:

$$\int \frac{\partial}{\partial k_y'} \delta(\vec{k}' - \vec{k}) A_{n,j'}(\vec{k}') \, d^3 k' = - \frac{\partial}{\partial k_y} A_{n,j'}(\vec{k}) \qquad (651)$$

$$\int \frac{\partial^2}{\partial k_y'^2} \delta(\vec{k}' - \vec{k}) A_{n,j'}(\vec{k}') \, d^3 k' = \frac{\partial^2}{\partial k_y^2} A_{n,j'}(\vec{k}) \qquad (652)$$

The coefficients $A_{n,j}(\vec{k})$ thus satisfy the differential equation

$$\sum_{j'} \left\{ \left[\frac{1}{2} \sum_{\alpha,\beta} k_\alpha k_\beta \frac{\partial^2 \varepsilon_n}{\partial k_\alpha \partial k_\beta} + iq \frac{B}{2\hbar} \frac{\partial^2 \varepsilon_n}{\partial k_y \partial k_x} \right] A_{n,j'}(\vec{k}) \right.$$

$$\left. + iq \frac{B}{\hbar} \sum_\alpha k_\alpha \frac{\partial^2 \varepsilon_n}{\partial k_\alpha \partial k_x} \frac{\partial}{\partial k_y} A_{n,j'}(\vec{k}) - \frac{q^2}{2\hbar^2} \frac{\partial^2 \varepsilon_n}{\partial k_x^2} \frac{\partial^2}{\partial k_y^2} A_{n,j'}(\vec{k}) \right\}$$

$$= \varepsilon A_{n,j}(\vec{k}) \tag{653}$$

To pass from the $\vec{k}$ wave vector space to real space, we must take account of the fact that

$$\hat{k}_\alpha = -i \frac{\partial}{\partial x_\alpha} \tag{503}$$

and make the transformation

$$A_{n,j'}(\vec{k}) \equiv \int F_{n,j'}(\vec{r}) \, e^{-i\vec{k}\cdot\vec{r}} \, d^3r \tag{654}$$

Equation (653) becomes

$$\sum_{j'} \left\{ \frac{1}{2} \sum_{\alpha,\beta} \hat{k}_\alpha \hat{k}_\beta \frac{\partial^2 \varepsilon_n}{\partial k_\alpha \partial k_\beta} + iq \frac{B}{2\hbar} \frac{\partial^2 \varepsilon_n}{\partial k_y \partial k_x} + \sum_\alpha q \frac{B}{\hbar} \frac{\partial^2 \varepsilon_n}{\partial k_\alpha \partial k_x} \hat{k}_\alpha \hat{y} \right.$$

$$\left. + \frac{q^2}{2\hbar^2} \frac{\partial^2 \varepsilon_n}{\partial k_x^2} \hat{y}^2 \right\} F_{n,j'}(\vec{r}) = \varepsilon F_{n,j}(\vec{r}) \tag{655}$$

In fact, (503) shows that

$$\sum_\alpha \hat{k}_\alpha \hat{y} = -i\hat{1} + \sum_\alpha \hat{y}\hat{k}_\alpha \tag{656}$$

an equation that we introduce in (655):

$$\sum_{j'} \left\{ \frac{1}{2} \sum_{\alpha,\beta} \hat{k}_\alpha \hat{k}_\beta \frac{\partial^2 \varepsilon_n}{\partial k_\alpha \partial k_\beta} + \frac{1}{2} q \frac{B}{\hbar} \sum_\alpha (\hat{k}_\alpha \hat{y} + \hat{y}\hat{k}_\alpha) \frac{\partial^2 \varepsilon_n}{\partial k_\alpha \partial k_x} \right.$$

$$\left. + q^2 \frac{B^2}{2\hbar^2} \hat{y}^2 \frac{\partial^2 \varepsilon_n}{\partial k_x^2} \right\} F_{n,j'}(\vec{r}) = \varepsilon F_{n,j}(\vec{r}) \tag{657}$$

which is also written in the form

$$\sum_{\substack{j' \\ \alpha,\beta}} \left\{ \frac{1}{2}\left[\hat{k}_\alpha + q\frac{B}{\hbar}\hat{y}\,\delta_{\alpha x}\right]\left[\hat{k}_\beta + q\frac{B}{\hbar}\hat{y}\,\delta_{\beta x}\right]\frac{\partial^2\varepsilon_n}{\partial k_\alpha \partial k_\beta}\right\} F_{n,j'}(\vec{r})$$

$$= \varepsilon F_{n,j}(\vec{r}) \tag{658}$$

This is the final result.

It is observed that if B = 0 (absence of magnetic field), (658) is written

$$\sum_{\substack{j' \\ \alpha,\beta}} \left\{ \frac{1}{2}\hat{k}_\alpha\hat{k}_\beta \frac{\partial^2\varepsilon_n}{\partial k_\alpha \partial k_\beta}\right\} F_{n,j'}(\vec{r}) = \varepsilon F_{n,j}(\vec{r}) \tag{659}$$

and hence if $B \neq 0$ it suffices in the different equations obtained in the case B = 0 to replace $\hat{k}_\alpha$ by

$$\hat{k}_\alpha + q\frac{B}{\hbar}\hat{y}\,\delta_{\alpha x} \tag{660}$$

That is, explicitly by applying a constant magnetic field parallel to the z axis, the following correspondences must be used for the components of the $\vec{k}$ vector:

$$\hat{k}_x \longrightarrow \hat{k}_x + q\frac{B}{\hbar}\hat{y}$$

$$\hat{k}_y \longrightarrow \hat{k}_y \tag{661}$$

$$\hat{k}_z \longrightarrow \hat{k}_z$$

To obtain Luttinger and Kohn's equation, it is necessary to introduce in the previous equations:

$$D_{\alpha\beta}^{jj'} = \sum_\alpha \frac{1}{2m}\,\delta_{\alpha\beta}\,\delta_{jj'} - \frac{1}{m^2}\sum_{\substack{n'',j'' \\ \alpha,\beta \\ n''\neq n}} \frac{p_\alpha^{\substack{n,j\ n'',j''}} p_\beta^{\substack{n'',j''\ n',j'}}}{\hbar\omega_{n''n}} \tag{662}$$

The critique of the calculations made is given in Ref. 53. We shall now show that if the matrix elements corresponding to the operator $\hat{\vec{p}}$ are not zero, the wave functions can also be written using the eigenfunctions of a harmonic oscillator.

XIV. HAMILTONIAN OF GRAPHITE IN THE PRESENCE OF A CONSTANT MAGNETIC FIELD [47,54-57]

A. Harmonic Oscillator

Let us briefly examine the movement of a unidimensional harmonic oscillator consisting of a helical spring with stiffness k. The differential equation governing the variation in its length x as a function of time is obtained by the fundamental principle of dynamics:

$$m\ddot{x} = -kx \tag{663}$$

called Hooke's law, where

$$k = mw^2 \tag{664}$$

The kinetic energy is given by

$$E_c = \frac{1}{2}m\dot{x}^2 \tag{665}$$

and the potential energy E_p by

$$\vec{F} = -\overrightarrow{\text{grad}}\, E_p \tag{666}$$

hence,

$$E_p = -\int_0^x F\,dx = \frac{1}{2}mw^2x^2 \tag{667}$$

if we assume for the origin of the potential energies the equilibrium position x = 0.

The definitions (451), (453), and (471) of the Lagrangian and Hamiltonian functions provide the equations

$$L = \frac{1}{2}m\dot{x}^2 - \frac{1}{2}mw^2x^2 \tag{668}$$

or

$$p = \frac{\partial L}{\partial \dot{x}} = m\dot{x} \tag{669}$$

and hence,

$$H = \frac{p^2}{2m} + \frac{1}{2} mw^2x^2 \tag{670}$$

If we pass on to the operators (application of Ehrenfest's theorems), since $\hat{x}$ and $\vec{\hat{p}}$ are related by the commutation equation

$$[\hat{x}, \ \vec{\hat{p}}] = i\hbar \tag{504}$$

and by setting

$$\hat{H} = \hbar w \hat{E} \tag{671}$$

$$\hat{x} = \sqrt{\frac{\hbar}{mw}} \ \hat{X} \tag{672}$$

$$\vec{\hat{p}} = \sqrt{m\hbar w} \ \vec{\hat{P}} \tag{673}$$

(the new quantities are dimensionless), we have in place of (670),

$$\hat{E} = \frac{1}{2} (\hat{X}^2 + \hat{P}^2) \tag{674}$$

By definition we shall denote $\hat{a}^+$ the so-called creation operator and $\hat{a}$ the annihilation operator such that

$$\hat{a} = \frac{1}{\sqrt{2}} (\hat{X} + i\hat{P}) \tag{675}$$

$$\hat{a}^+ = \frac{1}{\sqrt{2}} (\hat{X} - i\hat{P}) \tag{676}$$

since the operators $\hat{X}$ and $\hat{P}$ are Hermitian, $\hat{a}$ and $\hat{a}^+$ are Hermitian conjugates:

$$(\hat{a})^+ = \hat{a}^+ \tag{677}$$

$$(\hat{a}^+)^+ = \hat{a} \tag{678}$$

Equation (504) is transformed into

$$[\hat{X}, \ \hat{P}] = i \tag{679}$$

By inverting Eqs. (675) and (676), we have

$$\hat{X} = \frac{1}{\sqrt{2}} \, (\hat{a} + \hat{a}^+) \tag{680}$$

$$\hat{P} = -\frac{i}{\sqrt{2}} \, (\hat{a} - \hat{a}^+) \tag{681}$$

which we introduce into (679) to obtain

$$[\hat{a}, \hat{a}^+] = \hat{1} \tag{682}$$

but also, with (674),

$$\hat{E} = \frac{1}{2} \, (\hat{a}\hat{a}^+ + \hat{a}^+\hat{a}) \tag{683}$$

Now let

$$\hat{N} = \hat{a}^+\hat{a} \tag{684}$$

an operator which is Hermitian ($\hat{N}^+ = \hat{N}$); its eigenvalues are real. By using (682), we have

$$\hat{N}\hat{a} = \hat{a}^+\hat{a}\hat{a} = (\hat{a}\hat{a}^+ - \hat{1})\hat{a} = \hat{a}(\hat{N} - \hat{1}) \tag{685}$$

and

$$\hat{N}\hat{a}^+ = \hat{a}^+(\hat{1} + \hat{N}) \tag{686}$$

If we note $|n\rangle$ the eigenvector of $\hat{N}$ corresponding to the eigenvalue n, by definition,

$$\hat{N}|n\rangle = n|n\rangle \tag{687}$$

Since the norm of a vector is positive or zero, and since

$$\langle \hat{a}n|\hat{a}n\rangle = \langle n|\hat{a}^+\hat{a}n\rangle$$
$$= \langle n|\hat{N}n\rangle = n\langle n|n\rangle \tag{688}$$

$$\langle \hat{a}^+n|\hat{a}^+n\rangle = \langle n|\hat{a}\hat{a}^+n\rangle$$
$$= \langle n|(\hat{N} + \hat{1})n\rangle = (n + 1)\langle n|n\rangle \tag{689}$$

it is necessary to impose $n \geqslant 0$, where n is a pure number.

According to Eq. (685),

$$\hat{N}|\hat{a}n\rangle = \hat{a}(\hat{N} - \hat{1})|n\rangle = (n - 1)|\hat{a}n\rangle \tag{690}$$

implying that $\hat{a}|n\rangle$ is an eigenvector of the operator $\hat{N}$ for the eigenvalue n - 1. Hence $\hat{a}|n\rangle$ is proportional to $|n - 1\rangle$, which we denote

$$\hat{a}|n\rangle = \alpha_n|n - 1\rangle \tag{691}$$

Similarly, with (686),

$$\hat{N}|\hat{a}^+n\rangle = (n + 1)|\hat{a}^+n\rangle \tag{692}$$

namely,

$$\hat{a}^+|n\rangle = \beta_n|n + 1\rangle \tag{693}$$

It remains to evaluate α_n and β_n. If, by assumption, the eigenvectors $|n\rangle$ and $|n - 1\rangle$ are normed as well as $|n + 1\rangle$, according to (688),

$$\langle\hat{a}n|\hat{a}n\rangle = \alpha_n^2 = n \tag{694}$$

or

$$\alpha_n = \sqrt{n} \tag{695}$$

and

$$\hat{a}|n\rangle = \sqrt{n}|n - 1\rangle \tag{696}$$

In the same way, (689) serves to determine β_n. Thus,

$$\beta_n = \sqrt{n + 1} \tag{697}$$

and

$$\hat{a}^+|n\rangle = \sqrt{n + 1}|n + 1\rangle \tag{698}$$

Equations (682) through (684) give

$$\hat{E} = \hat{N} + \frac{\hat{1}}{2} \tag{699}$$

implying that the eigenvectors $|n\rangle$ of the operator $\hat{N}$ are also eigenvectors of the energy operator $\hat{E}$ whose eigenvalues are

$$E_n = n + \frac{1}{2} \tag{700}$$

where n = 0, 1, 2, 3, 4, ... Hence this quantum number assumes all the possible positive whole-number values.

B. Representation by Creation and
   Annihilation Operators [19-21]

The correspondence Eq. (660) established by Luttinger and Kohn involves the components of the $\vec{k}$ vector in a cartesian reference system. This forces us to recalculate the structural factors in such a location. By definition [22-24],

$$F_1(\vec{k}) = \sum_{i=1}^{6} e^{i\vec{k}\cdot\overrightarrow{A_0 A_i}} \tag{701}$$

and

$$F_2(\vec{k}) = \sum_{i=1}^{3} e^{i\vec{k}\cdot\overrightarrow{A_0 B_i}} \tag{702}$$

Since the origin is taken at A_0, Fig. 26 shows the position of the different atoms A_i and B_i.

The six first neighbors of A_0 of the type A are identified by

$$\overrightarrow{A_0 A_1} = \frac{a}{2}(-\sqrt{3}\,\vec{e}_1 + \vec{e}_2)$$

$$\overrightarrow{A_0 A_2} = \frac{a}{2}(-\sqrt{3}\,\vec{e}_1 - \vec{e}_2)$$

$$\overrightarrow{A_0 A_3} = a\vec{e}_2$$

$$\overrightarrow{A_0 A_4} = a\vec{e}_2 \tag{703}$$

$$\overrightarrow{A_0 A_5} = \frac{a}{2}(\sqrt{3}\,\vec{e}_1 - \vec{e}_2)$$

$$\overrightarrow{A_0 A_6} = \frac{a}{2}(\sqrt{3}\,\vec{e}_1 + \vec{e}_2)$$

and the three first neighbors of type B

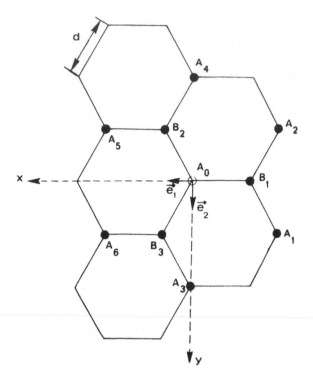

FIG. 26 First neighbors of a carbon atom of graphite:
a = d√3 = 2.46 Å.

$$\overrightarrow{A_0B_1} = -\frac{a}{\sqrt{3}}\,\vec{e}_1$$

$$\overrightarrow{A_0B_2} = \frac{a}{2}\left(\frac{\vec{e}_1}{\sqrt{3}} - \vec{e}_2\right) \tag{704}$$

$$\overrightarrow{A_0B_3} = \frac{a}{2}\left(\frac{\vec{e}_1}{\sqrt{3}} + \vec{e}_2\right)$$

Note that if $\vec{a}_i$ is a base vector of the space lattice and $\vec{b}_j$ a base vector of the reciprocal lattice,

$$\vec{a}_i \cdot \vec{b}_j = 2\pi\,\delta_{ij} \tag{705}$$

The explicit calculation of the scalar products involved in $F_1(\vec{k})$ and $F_2(\vec{k})$ and the use of Moivre's formula,

$$e^{ix} = \cos x + i \sin x \tag{706}$$

lead to the results

$$F_1(\vec{k}) = 4 \cos \pi a\sqrt{3} \, k_x \cos \pi ak_y + 2 \cos 2\pi k_y a \tag{707}$$

$$F_2(\vec{k}) = e^{-2\pi iak_x/\sqrt{3}} + 2 \cos \pi ak_y e^{i\pi ak_x/\sqrt{3}} \tag{708}$$

In the vicinity of the corner of the Brillouin zone, since

$$\vec{\Gamma P} = \vec{\Gamma K} + \vec{KP}$$

we can write

$$k_x = (k_x)_K + k'_x$$
$$k_y = (k_y)_K + k'_y \tag{709}$$

with

$$(k_x)_K = \frac{1}{\sqrt{3}a} \qquad \text{and} \qquad (k_y)_K = \frac{1}{3a} \tag{710}$$

These are the coordinates of point K counted from the origin Γ of the zone (Fig. 19) in the cartesian system of the reciprocal lattice constructed from that used in Fig. 5 for the space lattice.

Since in the neighborhood of the corner k'_x and k'_y are infinitely small, we can easily write the limited developments in the neighborhood of K of the functions $F_1(\vec{k})$ and $F_2(\vec{k})$. This yields

$$F_1(\vec{k}) = -3 + 3\pi^2 a^2 (k'^2_x + k'^2_y) \tag{711}$$

We remain with the first order in $\vec{k}$, so that the value of $F_1(\vec{k})$ used is independent of $\vec{k}$ and is -3.

The calculation is slightly longer for $F_2(\vec{k})$, but in the first order in $\vec{k}$, we obtain

$$F_2(\vec{k}) = -\sqrt{3}\pi ae^{i\pi/3} (k'_y - ik'_x) \tag{712}$$

Let us rewrite the matrix elements of the Hamiltonian operator $\hat{H}'$ in the absence of a magnetic field [formula (385)]:

$$(H'_{ij}) = \begin{pmatrix} H_{11} & H_{12} & H_{13} & H_{14} \\ H_{21} & H_{22} & H_{23} & H_{24} \\ H_{31} & H_{32} & H_{33} & H_{34} \\ H_{41} & H_{42} & H_{43} & H_{44} \end{pmatrix} \tag{713}$$

with Eqs. (386) and (351),

$$H_{11} = E_1^0 = \Delta + 2\gamma_1 \cos \psi + 2\gamma_5 \cos^2 \psi \tag{714}$$

Eq. (387) and

$$H_{12} = H_{21} = 0 \tag{715}$$

Eqs. (388) and (389),

$$H_{13} = \frac{1}{\sqrt{2}} (2\gamma_4 \cos \psi - \gamma_0) F_2(\vec{k}) = H_{14}^* \tag{716}$$

Eqs. (390) and (352),

$$H_{22} = E_2^0 = \Delta - 2\gamma_1 \cos \psi + 2\gamma_5 \cos^2 \psi \tag{717}$$

Eqs. (391) and (392),

$$H_{23} = \frac{1}{\sqrt{2}} (\gamma_0 + 2\gamma_4 \cos \psi) F_2(\vec{k}) = H_{24}^* \tag{718}$$

Eqs. (393) and (383),

$$H_{34} = 2\gamma_3 \cos \psi \, F_2(\vec{k}) \tag{719}$$

Eqs. (394) and (353),

$$H_{44} = H_{33} = E_3^0 = 2\gamma_2 \cos^2 \psi \tag{720}$$

if

$$\psi = k_z \frac{c}{2} \tag{721}$$

To simplify the resolution of the scalar determinant corresponding to the case of a nonzero constant magnetic field, it is necessary to set

$$\gamma_3 = \gamma_4 = 0 \tag{722}$$

and then to treat the case $\gamma_3 \neq 0$ by a perturbation method.

In the neighborhood of the corner K of the Brillouin zone, the action of a constant magnetic field is reflected by the replacement of $F_2(\vec{k})$ in Eqs. (714) to (721) by the value given in (712), and by the replacement of the k_α components of the $\vec{k}$ vector by

$$k_\alpha + \frac{qB}{\hbar} y \, \delta_{\alpha x}$$

with $\alpha = x, y, z$. We can therefore determine the correspondences of the operators [35]

$$\hat{k}'_x \equiv -i \frac{\partial}{\partial x} + \frac{q}{\hbar} By \tag{723}$$

$$\hat{k}'_y \equiv -i \frac{\partial}{\partial y} \tag{724}$$

$$\hat{k}'_z \equiv -i \frac{\partial}{\partial z} \tag{725}$$

If, with the help of (723) and (724), we form the commutator of $\hat{k}'_x$ and $\hat{k}'_y$, this gives

$$[\hat{k}'_x, \hat{k}'_y] = \frac{iqB}{\hbar} \tag{726}$$

In agreement with McClure, let

$$s = \frac{qB}{\hbar} \tag{727}$$

Defining the creation and annihilation operators

$$\hat{k}^+ = -i \frac{e^{i(\pi/3)}}{\sqrt{2}} (\hat{k}'_y - i\hat{k}'_x) \tag{728}$$

and

$$\hat{k} = i \frac{e^{-i(\pi/3)}}{\sqrt{2}} (\hat{k}'_y + i\hat{k}'_x) \tag{729}$$

by analogy with $\hat{a}$ and $\hat{a}^+$ [Eqs. (675) and (676)].

By taking account of the fact that a factor s appears in the
commutator (726), which does not exist in (679), we retranscribe
(696) and (698) in the form

$$\hat{k}|n\rangle = \sqrt{ns}\,|n-1\rangle \tag{730}$$

$$\hat{k}^+|n\rangle = \sqrt{(n+1)s}\,|n+1\rangle \tag{731}$$

where the ket $|n\rangle$ is the eigenstate of a harmonic oscillator equiva-
lent to the system $n = 0, 1, 2, 3, \ldots$ its quantum number

C. McClure's Method

With these different notations, we are led to use the operators

$$\hat{F}_2(\hat{k}) = -i\sqrt{6}\pi a\hat{k}^+ \tag{732}$$

[Eqs. (712) and (728)]

$$(\hat{F}_2(\hat{k}))^* = i\sqrt{6}\pi a\hat{k} \tag{733}$$

which, introduced in (713), give

$$(H'_{ij}) = \begin{pmatrix} E_1^0 & 0 & i\sqrt{3}\pi a\gamma_0\hat{k}^+ & -i\sqrt{3}\pi a\gamma_0\hat{k} \\ 0 & E_2^0 & -i\sqrt{3}\pi a\gamma_0\hat{k}^+ & -i\sqrt{3}\pi a\gamma_0\hat{k} \\ -i\sqrt{3}\pi a\gamma_0\hat{k} & i\sqrt{3}\pi a\gamma_0\hat{k} & E_3^0 & 0 \\ i\sqrt{3}\pi a\gamma_0\hat{k}^+ & i\sqrt{3}\pi a\gamma_0\hat{k}^+ & 0 & E_3^0 \end{pmatrix} \tag{734}$$

provided that (722) is satisfied.

To use McClure's notation for (734), it is necessary to set

$$\hbar v = \sqrt{3}\pi a\gamma_0 \tag{735}$$

[the factor 2π that differentiates Eq. (735) from McClure's convention derives from the choice of unitary vectors, that is, from the
writing of the scalar product (705)].

To obtain the energy bands in the presence of a constant magnetic field, the eigenvalue equation

$$H'|\Psi\rangle = \varepsilon|\Psi\rangle \tag{736}$$

must be resolved. The structure of (H'_{ij}) given by (734) leads us to assume for Ψ a column vector of which the four components are $a_1|n\rangle$, $a_2|n\rangle$, $a_3|n - 1\rangle$, and $a_4|n + 1\rangle$, where the a_α (α = 1, 2, 3, 4) are coefficients to be determined.

In fact, by using the properties (728) and (729), the secular determinant (736) is transformed into the resolution of

$$
\begin{pmatrix}
E_1^0 & 0 & i\pi a\gamma_0\sqrt{3n}s & -i\pi a\gamma_0\sqrt{3(n+1)}s \\
0 & E_2^0 & -i\pi a\gamma_0\sqrt{3n}s & -i\pi a\gamma_0\sqrt{3(n+1)}s \\
-i\pi a\gamma_0\sqrt{3n}s & i\pi a\gamma_0\sqrt{3n}s & E_3^0 & 0 \\
i\pi a\gamma_0\sqrt{3(n+1)}s & i\pi a\gamma_0\sqrt{3(n+1)}s & 0 & E_3^0
\end{pmatrix}
\begin{pmatrix} a_1 \\ a_2 \\ a_3 \\ a_4 \end{pmatrix}
= \varepsilon
\begin{pmatrix} a_1 \\ a_2 \\ a_3 \\ a_4 \end{pmatrix}
$$

(737)

which allows nonzero solutions for a_α if, for $n \geqslant 1$, the energy ε satisfies

$$
(E_1^0 - \varepsilon)(E_2^0 - \varepsilon)(E_3^0 - \varepsilon)^2 - 3\pi^2 a^2 \gamma_0^2 (2n + 1)s(E_1^0 - \varepsilon)(E_3^0 - \varepsilon)
$$

$$
- 3\pi^2 a^2 \gamma_0^2 (2n + 1)s(E_2^0 - \varepsilon)(E_3^0 - \varepsilon) + 36\pi^4 a^4 \gamma_0^4 n(n + 1)s^2 = 0
$$

(738)

replacing (397). It differs from the latter in that it cannot be placed in the form of a product of two equations of the second degree. The magnetic field mixes the states of the different bands:

$$
[(E_1^0 - \varepsilon)(E_3^0 - \varepsilon) - 3\pi^2 a^2 \gamma_0^2 (2n + 1)s][E_2^0 - \varepsilon)(E_3^0 - \varepsilon)
$$

$$
- 3\pi^2 a^2 \gamma_0^2 (2n + 1)s] - 9\pi^4 a^4 \gamma_0^4 s^2 = 0
$$

(739)

The term $9\pi^4 a^4 \gamma_0^4 s^2$ corresponds to the interaction between the different bands in the presence of the constant magnetic field. If we ignore it, the semiclassic energy spectrum is obtained:

$$
\varepsilon_{n_{1\pm}} = \frac{1}{2}(E_1^0 + E_3^0) \pm \left[\frac{1}{4}(E_1^0 - E_3^0)^2 + 3\pi^2 a^2 \gamma_0^2 (2n + 1)q\frac{B}{\hbar}\right]^{1/2}
$$

(740)

$$
\varepsilon_{n_{2\pm}} = \frac{1}{2}(E_2^0 + E_3^0) \pm \left[\frac{1}{4}(E_2^0 - E_3^0)^2 + 3\pi^2 a^2 \gamma_0^2 (2n + 1)q\frac{B}{\hbar}\right]^{1/2}
$$

(741)

In Ref. 60, McClure shows that the term ignored, $9\pi^4 a^4 \gamma_0^2 s^2$, has very
little influence for $k_z = 0$ on $\varepsilon_{n_{1+}}$ and $\varepsilon_{n_{2-}}$ but may make a considerable contribution for the bands $\varepsilon_{n_{1-}}$ and $\varepsilon_{n_{2+}}$. A perturbation
method also helps to find the solutions for $\gamma_3 \neq 0$ [32].

GRAPHITE: DENSITY OF STATES
XV. FERMI-DIRAC STATISTIC [58]
A. Distribution Function

The theoretical study of graphite that we have carried out is
actually that of the delocalized π electrons. These are identical,
indistinguishable particles forming a system for which only the
antisymmetrical states are accessible: it is necessary to satisfy
Pauli's exclusion principle. Each quantum state can only contain
zero or one particle.

A microscopic state is defined by the data of N_1, N_2, . . ., N_i,
where N_i is the number of electrons occupying energy level ε_i. Each
of these levels is characterized by its degeneration g_i (where $g_i \gg 1$
and $N_i \leqslant g_i$ due to Pauli's principle).

The number of complexions or microscopic states corresponding
to a given macroscopic state can be evaluated as follows. Since the
particles are indistinguishable, there is only one way to distribute
N electrons over the energy levels by placing N_1 on level ε_1, N_2 on
level ε_2, and so on.

The number of microscopic states is thus obtained by calculating
the distribution on the energy sublevels. For an energy ε_i, we have
a distribution of N_i electrons on g_i sublevels, which are occupied or
vacant, and if we wish to obtain a new distribution it is necessary
to interchange the vacant spaces and the occupied space, giving $g_i!$
possible permutations. However, if two vacant spaces [$(g_i - N_i)!$
possibilities] or two occupied spaces ($N_i!$ possibilities) are permuted,
a new distribution is not obtained. Hence,

$$\Omega_i = \frac{g_i!}{N_i!(g_i - N_i)!} \tag{742}$$

and the number of ways in which to distribute N indistinguishable particles on levels ε_1, ε_2, and so on, by placing N_1, N_2, and so forth, electrons respectively, on these levels is

$$\Omega = \prod_i \frac{g_i!}{N_i!(g_i - N_i)!} \tag{743}$$

Stirling's formula,

$$\log x! = x \log x - x \tag{744}$$

applied to (743) yields

$$\log \Omega = \sum_i \left\{ N_i \log \left(\frac{g_i - N_i}{N_i} \right) - g_i \log \left(\frac{g_i - N_i}{g_i} \right) \right\} \tag{745}$$

The thermodynamic probability is maximum if $d\Omega$ or, what amounts to the same thing, $d(\log \Omega)$ is zero:

$$\sum_i \log \left(\frac{g_i}{N_i} - 1 \right) dN_i = 0 \tag{746}$$

By accounting for the fact that the total number of particles N and the total energy E are constants and that

$$E = \sum_i N_i \varepsilon_i \tag{747}$$

$$N = \sum_i N_i \tag{748}$$

it is necessary to add to (746) the conditions

$$\sum_i \varepsilon_i \, dN_i = 0 \tag{749}$$

$$\sum_i dN_i = 0 \tag{750}$$

If λ_1 and λ_2 are Lagrangian multipliers [43], by forming

$$(746) - \lambda_1(750) - \lambda_2(749) = 0$$

we obtain

$$N_i = \frac{g_i}{e^{\lambda_1 + \lambda_2 \varepsilon_i} + 1} \tag{751}$$

which expresses the electron distribution at equilibrium. It is shown that λ_2 is independent of the statistic used, but in the Maxwell-Boltzmann statistics,

$$\lambda_2 = \frac{1}{kT} \tag{752}$$

where $k = 1.381 \times 10^{-23}$ J deg^{-1} = 8.62×10^{-5} eV deg^{-1}. To simplify the writing, we set

$$\lambda_1 = -\frac{\varepsilon_F}{kT} \tag{753}$$

and

$$f(\varepsilon, T) = \frac{N_i}{g_i} \tag{754}$$

hence,

$$f(\varepsilon, T) = \frac{1}{1 + e^{(\varepsilon - \varepsilon_F)/kT}} \tag{755}$$

This is the Fermi-Dirac distribution rule. It is interesting to plot this curve using the temperature T as a parameter. The origin O (see Fig. 21) is the center of symmetry. The function $f(\varepsilon)$ is always decreasing because

$$\frac{\partial f}{\partial \varepsilon} = -\frac{1}{kT} \frac{1}{[e^{-(\varepsilon - \varepsilon_F)/kT} + e^{(\varepsilon - \varepsilon_F)/kT}]^2} \tag{756}$$

The second derivative is nullified for $\varepsilon = \varepsilon_F$ and is an inflection point.

B. Fermi Energy [59]

Note that if

$$\varepsilon \longrightarrow -\infty \qquad f(\varepsilon) \longrightarrow 1$$

$$\varepsilon \longrightarrow +\infty \qquad f(\varepsilon) \longrightarrow 0$$

and assuming T = 0 K if

$$\varepsilon < \varepsilon_F \qquad f(\varepsilon) \longrightarrow 1$$

$$\varepsilon > \varepsilon_F \qquad f(\varepsilon) \longrightarrow 0$$

that at zero absolute all the energy levels are occupied up to a value $\varepsilon = \varepsilon_F$ and all the energy levels are vacant for $\varepsilon > \varepsilon_F$. This value ε_F is called the Fermi energy.

By increasing the temperature a number of electrons begin to occupy energy levels higher than ε_F. The Fermi energy is hence the dividing line between the energy levels occupied by the electrons and the levels that remain vacant at the temperature 0 K. It corresponds to the Gibbs thermodynamic potential, that is, to the work that must be supplied to increase by 1 the number of particles of the system investigated.

Whatever the temperature $T \neq 0$ K, for $\varepsilon = \varepsilon_F$ the probability $f(\varepsilon)$ is 0.5. Roughly, if

$$\varepsilon \ll \varepsilon_F \qquad f(\varepsilon) \sim 1$$

$$\varepsilon \gg \varepsilon_F \qquad f(\varepsilon) \simeq e^{-\varepsilon/kT}$$

we return to the Maxwell-Boltzmann statistics. An interesting and useful remark may be made. If we set

$$\xi = \frac{\varepsilon - \varepsilon_F}{kT}$$

then Eq. (755) becomes

$$f = (1 + e^{\xi})^{-1}$$

and, for $-2 \leqslant \xi \leqslant 2$, the numerical calculation of f gives $0.88 \geqslant f \geqslant 0.12$, implying that, to within 12%, the entire variation of $f(\varepsilon)$ occurs in a 2kT energy band surrounding the Fermi energy ε_F.

For crystals, Kronig and Penney [18] have demonstrated that the periodicity of the potential causes the existence of permitted and prohibited energy bands (Fig. 27). When an energy band is vacant, it obviously makes no contribution to the conduction of

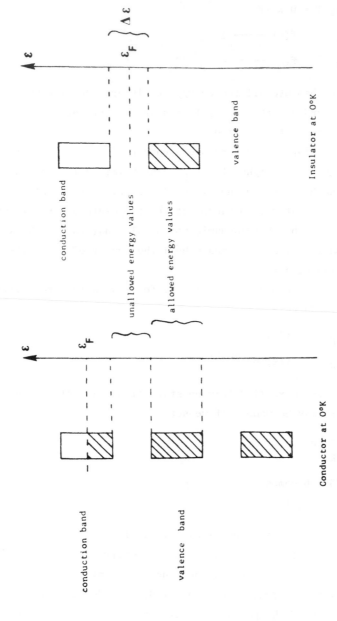

FIG. 27 Conductors and insulators.

graphite on the application of an electric field. If the energy
band is full, all the energy levels are occupied by electrons, but
owing to Pauli's exclusion principle, by the application of an
electric field no ordered movement of all the electrons can take
place. The electrons of a crystal are potential conducting electrons only if they belong to incompletely filled energy bands. The
absence of electrons at the top of an energy band is equivalent to
the presence of mobile, fictitious particles with charge +e, called
holes.

If the Fermi energy lies within a permitted band, the crystal
is conducting if an external electric field is applied. The energy
band containing ε_F is the so-called conduction band, and the band
immediately below it is the valence band. In the case of an insu-
lator ($\Delta\varepsilon > 5$ eV) or a semiconductor ($\Delta\varepsilon \simeq 1$ eV), the Fermi energy
lies in a prohibited band. Thus the expression "valence band" is
applied to the permitted energy band lying immediately below ε_F and
the conduction band is that lying immediately above ε_Γ.

For graphite at 0 K, the hexagonal Brillouin zone is full and
the adjacent zone is completely vacant (see Secs. I-IV). According
to Fig. 27, the Fermi energy is the energy of the highest occupied
level, $\varepsilon_F = \varepsilon_K$, if K represents the corner of the Brillouin zone.
This means that the band E_- corresponds to the interior of the zone
and E_+ to the exterior if E_- and E_+ are the solutions of the secular
determinant of plane graphite.

XVI. DISPERSION RULE

A. Absence of a Magnetic Field

In the case of graphite, and assuming as a basis for discussion the
plane model largely developed in Refs. 29, 30, 33, 34, 54, and 55
and in this chapter, the tight binding approximation leads to a
system with two energy bands:

$$E_\pm = E_0 \pm \gamma_0 |F_2(\vec{k})| \pm \gamma_0' F_1(\vec{k}) \mp \Delta_A \tag{757}$$

In this equation, E_0 is the energy of an electron $2P_z$ for an isolated carbon atom, and γ_0, γ_0', and Δ_A are the interaction parameters between the different carbon atoms (Fig. 15):

$$\gamma_0 = <\phi(\vec{r} - \vec{R}_A)|\hat{H}'|\phi(\vec{r} - \vec{R}_B)> \tag{758}$$

$$\gamma_0' = <\phi(\vec{r} - \vec{R}_A)|\hat{H}'|\phi(\vec{r} - \vec{R}_{A'})> \tag{759}$$

$$\Delta_A = -<\phi(\vec{r} - \vec{R}_A)|\hat{H}'|\phi(\vec{r} - \vec{R}_A)> \tag{760}$$

In the discussion below, we also need the following:

$$\Delta_B = -<\phi(\vec{r} - \vec{R}_B)|\hat{H}'|\phi(\vec{r} - \vec{R}_B)> \tag{761}$$

$$\gamma_1 = <\phi(\vec{r} - \vec{R}_A)|\hat{H}'|\phi(\vec{r} - \vec{R}_C)> \tag{762}$$

$$\gamma_2 = 2<\phi(\vec{r} - \vec{R}_B)|\hat{H}'|\phi(\vec{r} - \vec{R}_B - \vec{a}_3)> \tag{763}$$

$$\gamma_5 = 2<\phi(\vec{r} - \vec{R}_A)|\hat{H}'|\phi(\vec{r} - \vec{R}_A - \vec{a}_3)> \tag{764}$$

where $\vec{a}_3$ is the translation vector of the space lattice in a perpendicular direction to the graphite plane. The quantity H' may be considered a perturbation:

$$\hat{H}' = \hat{V}(\vec{r}) - \hat{U}(\vec{r} - \vec{R}_i) \tag{765}$$

It is negative. $\hat{U}(\vec{r} - \vec{R}_i)$ is the potential energy of a $2P_z$ electron in an isolated carbon atom, and $\hat{V}(\vec{r})$ is the potential energy of a π electron in the graphite lattice. The wave functions $\phi(\vec{r} - \vec{R}_i)$ are Bloch's functions associated with families of atoms (A, B, C, and D) making up the graphite.

$F_1(\vec{k})$ and $F_2(\vec{k})$ are the structural factors; hence by restricting ourselves to the closest neighbors (see Secs. VIII-IX),

$$F_1(\vec{k}) = \sum_{i=1}^{6} e^{i\vec{k}\cdot\overrightarrow{A_0A_i}} = 4 \cos \pi a\sqrt{3}k_x \cos \pi ak_y + 2 \cos 2\pi k_y a \tag{766}$$

$$F_2(\vec{k}) = \sum_{i=1}^{3} e^{i\vec{k}\cdot\overrightarrow{A_0B_i}} = e^{-(2\pi ia/\sqrt{3})k_x} + 2 \cos \pi ak_y e^{(i\pi a/\sqrt{3})k_x} \tag{767}$$

In the absence of a magnetic field and in the neighborhood of the corner K of the Brillouin zone (Fig. 19), at the first order in $\vec{k}$,

$$F_1(\vec{k}) = -3 \tag{768}$$

$$F_2(\vec{k}) = -\sqrt{3}\pi a e^{i(\pi/3)}(k_y' - ik_x') \tag{769}$$

k_x' and k_y' are the components of the $\vec{k}$ vector counted from point K of the Brillouin zone taken as the origin of a cartesian reference system of the reciprocal lattice. The constant a is 2.46 Å.

By convention, the energies are measured in relation to that of point K. Inside the Brillouin zone,

$$E_K = E_0 + 3\gamma_0' + \Delta_A \tag{770}$$

and at a point M very near point K, the energy of the band is written

$$E_- = -\gamma_0 |F_2(\vec{k})| = \sqrt{3}\pi a \gamma_0 \sqrt{k_x'^2 + k_y'^2} \tag{771}$$

that we denote, since

$$\vec{KM} = \vec{k} = \vec{i}k_x' + \vec{j}k_y' \tag{772}$$

$$E_- = \sqrt{3}\pi a \gamma_0 |\vec{k}| - \hbar v |\vec{k}| \tag{773}$$

The notation used is that of McClure [30,33]:

$$\hbar v = \sqrt{3}\pi a \gamma_0 \tag{774}$$

Even if the approximation of the plane model must be corrected, one can thus draw the conclusion that the π electrons of graphite do not behave like the free electrons. In fact, for the free electrons, that is, electrons not subjected to forces deriving from a potential, the total mechanical energy is purely kinetic:

$$E = \frac{p^2}{2m} = \frac{\hbar^2 k^2}{2m} \tag{775}$$

The dispersion rule is parabolic.

For graphite, the relationship between energy and the wave vector is linear, justifying, a posteriori, ignoring the k^2 terms in the different limited developments made until now.

If we do not ignore the interactions between the different graphite planes, the three-dimensional model obtained by means of the tight binding approximation is that developed in the previous sections and is called the Slonczewski-Weiss model.

The energy bands are as follows, taking the energy at point H of the Brillouin zone as the origin (Fig. 19); that is,

$$E_0 - \Delta_B - \gamma_2 = 0 \tag{776}$$

by setting

$$\psi = k_z \frac{c}{2} \tag{777}$$

with

$$-\frac{\pi}{c} \leqslant k_z \leqslant +\frac{\pi}{c} \tag{778}$$

and

$$\Delta = \Delta_B - \Delta_A + \gamma_2 - \gamma_5 \tag{779}$$

(this quantity Δ is extremely small [60]). The parameters Δ_B, γ_2, γ_5, and so on, have been defined in Eqs. (758) through (764).

$$E_1^0 = \Delta + 2\gamma_1 \cos \psi + 2\gamma_5 \cos^2 \psi \tag{780}$$

$$E_2^0 = \Delta - 2\gamma_1 \cos \psi + 2\gamma_5 \cos^2 \psi \tag{781}$$

$$E_3^0 = E_4^0 = 2\gamma_2 \cos^2 \psi \tag{782}$$

The last band is doubly degenerate. Figure 28 offers an illustration of this. The lines have been plotted along the edge HKH of the Brillouin zone, that is, by varying k_z from $-\pi/c$ to $+\pi/c$. The order of magnitude of the interaction parameters [60] enables us to restrict ourselves to the functions

$$E_1^0 = 2\gamma_1 \cos k_z \frac{c}{2} \tag{783}$$

$$E_2^0 = -2\gamma_1 \cos k_z \frac{c}{2} \tag{784}$$

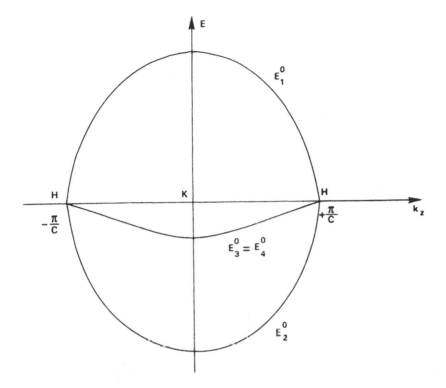

FIG. 28   Graphite band model.

$$E_3^0 = E_4^0 = 2\gamma_2 \cos^2 k_z \frac{c}{2} \tag{785}$$

It should be noted that for the energy levels near E_3^0, a ternary symmetry exists (trigonal warping).

B.   Landau Levels

In the presence of a constant magnetic field, the Landau levels in a semiclassic first approximation are given by the four bands

$$\varepsilon_{n_{1\pm}} = \frac{1}{2}(E_1^0 + E_3^0) \pm \left[\frac{1}{4}(E_1^0 - E_3^0)^2 + (\hbar v)^2(2n + 1)s\right]^{1/2} \tag{786}$$

$$\varepsilon_{n_{2\pm}} = \frac{1}{2}(E_2^0 + E_3^0) \pm \left[\frac{1}{4}(E_2^0 - E_3^0)^2 + (\hbar v)^2(2n + 1)s\right]^{1/2} \tag{787}$$

if

$$s = \frac{qB}{\hbar} \qquad\qquad\qquad\qquad (788)$$

They are valid in the neighborhood of the edge HKH of the Brillouin zone. If we wish to represent them schematically as a function of $|\vec{K}|$ for constant k_z (Fig. 29), it is first necessary to relate (786) and (787) to the results of Sugihara [61,62], who used a $\vec{k}\cdot\vec{p}$ method.

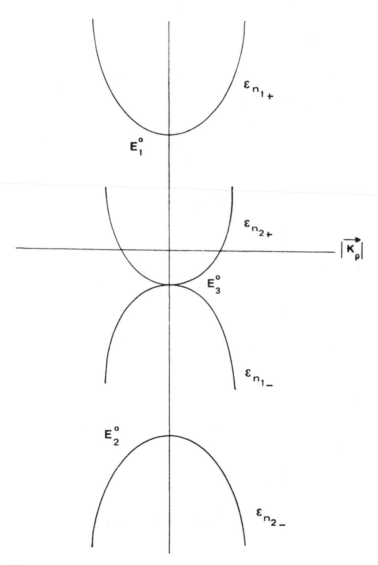

FIG. 29   Landau levels.

During examination of the magnetic Hamiltonian, we showed that for a harmonic oscillator the dimensionless Hamiltonian operator is

$$\hat{E} = \frac{1}{2}(\hat{a}\hat{a}^+ + \hat{a}^+\hat{a}) \tag{789}$$

if

$$\hat{a} = \frac{1}{\sqrt{2}}(\hat{X} + i\hat{P}) \tag{790}$$

and

$$[\hat{X}, \hat{P}] = i \tag{791}$$

McClure's method caused us to set

$$\hat{k} = \frac{ie^{-i\pi/3}}{\sqrt{2}} \ (\hat{k}'_y + i\hat{k}'_x) \tag{792}$$

with

$$[\hat{k}'_x, \hat{k}'_y] = is \tag{793}$$

The equivalent of (789) is hence

$$\hat{E} = \frac{1}{2}(\hat{k}\hat{k}^+ + \hat{k}^+\hat{k}) \tag{794}$$

or, using definition (792),

$$\hat{E} = \frac{1}{2}(\hat{k}'^2_x + \hat{k}'^2_y) \tag{795}$$

and the eigenvalues of the operator $\hat{E}$ are written

$$E_n = (n + \frac{1}{2})s \tag{796}$$

to take account of the s factor, which is involved in (793), not in (791). This implies the equivalence, at constant k_z, between $(n + 1/2)s$ and $1/2(k'^2_x + k'^2_y)$ and Sugihara's formulas

$$\varepsilon_{n_{1\pm}} = \frac{1}{2}(E_1^0 + E_3^0) \pm \left[\frac{1}{4}(E_1^0 - E_3^0)^2 + (\hbar v)^2(k'^2_x + k'^2_y)\right]^{1/2} \tag{797}$$

$$\varepsilon_{n_{2\pm}} = \frac{1}{2}(E_2^0 + E_3^0) \pm \left[\frac{1}{4}(E_2^0 - E_3^0)^2 + (\hbar v)^2(k'^2_x + k'^2_y)\right]^{1/2} \tag{798}$$

To obtain Fig. 29, it suffices to take

$$k_\rho = \sqrt{{k'_x}^2 + {k'_y}^2}$$

as a variable. Figure 19 offers a better understanding of this notation.

XVII. DENSITY OF STATES [15,16]

We have shown that if Ω is the volume constructed using the base vectors of the space lattice (unitary cell of the Bravais lattice), a volume

$$\Omega_r = \frac{8\pi^3}{\Omega} \tag{799}$$

corresponds to it in the reciprocal lattice. If N is the number of unitary cells, the Brillouin zone contains exactly N possible $\vec{k}$ values and the number of states per element of volume d^3k is thus

$$\frac{N\Omega}{8\pi^3} \, d^3k \tag{800}$$

(if spin is ignored). By definition, the function N(E), called the density of states, is equal to the number of permitted states counted in the volume of the crystal for a unit energy interval. Consider that in the reciprocal space, that is, in the Brillouin zone, the isoenergetic surfaces σ and σ' correspond to the value ε and $\varepsilon + d\varepsilon$ of the energy, respectively (Fig. 30).

Let us denote $d\sigma$ a surface element of σ and k_n the projection of the $\vec{k}$ wave vector on the element $\vec{n}$ normal to this surface.

The number of states between σ and σ', hence lying between the energy values ε and $\varepsilon + d\varepsilon$, is equal to $2N\Omega/8\pi^3$ times the volume lying between the surfaces σ and σ'. The factor 2 derives from the electron spin. A volume element d^3k may be written in the form $dk_n \, d\sigma$. By definition of the gradient vector of a function,

$$d\varepsilon = \overrightarrow{\mathrm{grad}_k}\varepsilon \cdot \overrightarrow{dk} \tag{801}$$

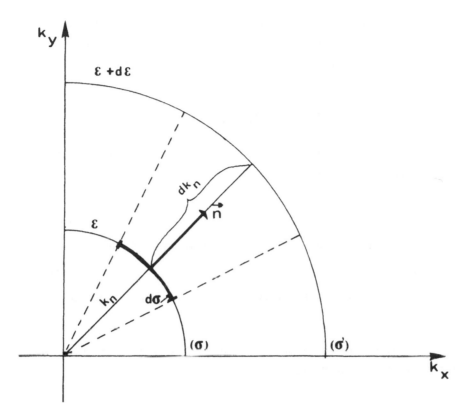

FIG. 30   Isoenergy surfaces and density of states.

the vector $\vec{\text{grad}}\ \varepsilon$ is perpendicular to the isoenergetic surface σ and hence parallel to the normal $\vec{n}$:

$$dk_n = \frac{d\varepsilon}{|\vec{\nabla}_k \varepsilon|} \tag{802}$$

The density of states is thus

$$N(\varepsilon)d\varepsilon = \frac{2N\Omega}{8\pi^3} \int_\sigma \frac{d\sigma d\varepsilon}{|\vec{\nabla}_k \varepsilon|} \tag{803}$$

and the number of states per unit energy,

$$N(\varepsilon) = \frac{2V}{8\pi^3} \int_\sigma \frac{d\sigma}{|\vec{\nabla}_k \varepsilon|} \tag{804}$$

where V is the total volume of the crystal. Integration is carried out on an isoenergy curve σ. Equation (84) serves to evaluate, at thermodynamic equilibrium, the number of electrons dn with energy lying between ε and ε + dε:

$$dn = N(\varepsilon) f(\varepsilon) \; d\varepsilon = \frac{N(\varepsilon) \; d\varepsilon}{1 + e^{(\varepsilon - \varepsilon_F)/kT}} \tag{805}$$

Let us apply the definition (84) in the case of graphite. In the neighborhood of point K, for the two-dimensional model, that is, in the neighborhood of the Fermi energy,

$$\varepsilon = \hbar v |\vec{K}| \tag{773}$$

and the isoenergetic surfaces are the circle loci of the points

$$|\vec{K}| = \text{constant} \tag{806}$$

Since our model is two-dimensional, σ is a line, not a surface:

$$\sigma = 2\pi |\vec{K}| \tag{807}$$

and it is necessary to replace $V/8\pi^3$ by $S/4\pi^2$, where S is the area of the crystal. According to (773),

$$|\vec{\nabla}_k \varepsilon| = \hbar v \tag{808}$$

thus,

$$N(\varepsilon) = \frac{S}{\pi (\hbar v)^2} |\varepsilon| \tag{809}$$

where ε represents the energy in relation to the corner, that is, taking the Fermi energy as zero. Curve 22 is an illustration of (809). According to the discussion in Sec. XV.B, we shall have holes for $\varepsilon < \varepsilon_F$ and electrons for $\varepsilon > \varepsilon_F$ when $T \neq 0$ K. The valence band and the conduction band touch at ε_F: as a first approximation, graphite is often considered a zero-gap semiconductor.

XVIII. FERMI LEVEL

We have shown the great importance of ε_F to distinguish between conduction by electrons and conduction by holes, which is possible by the application of an electric field.

In relation to the plane model, let us evaluate the value of ε_F and its variation as a function of temperature. We know that the density of the energy levels is proportional to $|\varepsilon|$:

$$N(\varepsilon) = \alpha |\varepsilon| \tag{810}$$

The effective number of carriers is the sum of the free electrons and holes, the distribution function corresponding to them is $f(\varepsilon)$ and $1 - f(\varepsilon)$, respectively (see Fig. 21), and (805) enables us to evaluate this effective number n_{eff}:

$$n_{eff} = \int_{-\infty}^{+\infty} dn = \int_{-\infty}^{0} \alpha |\varepsilon| (1 - f(\varepsilon)) \, d\varepsilon + \int_{0}^{\infty} \alpha |\varepsilon| f(\varepsilon) \, d\varepsilon \tag{811}$$

Note that at the temperature of 0 K, all the electron states are occupied ($f(\varepsilon) = 1$ for $\varepsilon < \varepsilon_F$ up to $\varepsilon = \varepsilon_F$, and if $\varepsilon \geqslant \varepsilon_F$, $f(\varepsilon) = 0$). Hence,

$$n_{eff} = \int_{-\infty}^{(\varepsilon_F)_0} \alpha |\varepsilon| \, d\varepsilon = \frac{\alpha}{2}(\varepsilon_F)_0^2 \tag{812}$$

The quantity $(\varepsilon_F)_0$ is the value of the Fermi energy at 0 K. To simplify the writing let us set

$$\varepsilon = ukT \tag{813}$$

$$\varepsilon_F = \theta kT \tag{814}$$

$$x = u - \theta \tag{815}$$

Equation (811) becomes

$$\frac{\theta_0^2}{2} = \theta \left[\int_{-\infty}^{-\theta} \frac{e^x \, dx}{1 + e^x} + \int_{-\theta}^{0} \frac{e^{-x} \, dx}{1 + e^{-x}} + \int_{0}^{\infty} \frac{e^{-x} \, dx}{1 + e^{-x}} \right] + \int_{-\infty}^{-\theta} \frac{xe^x}{1 + e^x} \, dx$$

$$+ \int_{-\theta}^{0} \frac{xe^{-x}}{1 + e^{-x}} \, dx + \int_{0}^{\infty} \frac{xe^{-x}}{1 + e^{-x}} \, dx \tag{816}$$

This leads to

$$\theta_0^2 = 2\theta \, \log \frac{1 + e^{\theta}}{1 + e^{-\theta}} - 3\theta^2 + 4\int_{0}^{\theta} \log \, (1 + e^x) \, dx \tag{817}$$

Since

$$\int \log (1 + e^x) \, dx = \int \log (1 + e^{-x}) \, dx + \frac{x^2}{2} \tag{818}$$

If we consider the case of high temperatures,

$$\theta \ll 1 \qquad \text{and} \qquad \theta_0^2 = \theta \log 16 \tag{819}$$

that is,

$$\varepsilon_F = \frac{(\varepsilon_F)_0^2}{kT \log 16} \tag{820}$$

For low temperatures, $\theta \gg 1$ and

$$\int_0^\infty \log (1 + e^{-x}) \, dx = \frac{\pi^2}{12} \tag{821}$$

$$\theta_0^2 = \theta^2 + \frac{\pi^2}{3}$$

Hence,

$$\varepsilon_F = \left[(\varepsilon_F)_0^2 - \frac{\pi^2}{3} (kT)^2 \right]^{1/2} \tag{822}$$

The variation in Fermi energy as a function of temperature has a major effect on the value of the magnetic susceptibility of graphite. Note also Ref. 63 to conclude that the understanding of susceptibility requires the knowledge of the density of states, but if transfer mechanisms are investigated, it is also necessary to determine the collision mechanisms and the influence of the magnetic field on these effects.

In conclusion, note that the experiment serves to clarify the value of the different parameters introduced in the calculation of the graphite bands. These results allow the interpretation of cyclotron resonance processes and magneto-optical effects. This study may be supplemented by reading the article by Nozières [64] and its criticism by Inoue. The parameters of Slonczewski and Weiss, γ_0, γ_1, and γ_2, can thus be determined experimentally.

This text merits a serious supplement dealing with the calculation of the numerical values, ab initio or with the help of experiments, of the parameters introduced by the band theory using the tight binding approximation. This chapter should be supplemented by the application of this model to the evaluation of the Fermi energy, the susceptibility of graphite, the interpretation of the Haas Van Alphen effect, of electrical conductivities, of Hall's coefficient, and of the transverse magnetic resistivity.

In *The Physics of Semi-Metals and Narrow-Gap Semiconductors,* edited by Carter and Bate and published by the Pergamon Press, Chapter III, Graphite, completes our discussion. The reader who has followed us to the end should be able to cope with this material without major problems. Also worth noting is the excellent Volume 8 of the *Chemistry and Physics of Carbon* series edited by Walker and Thrower and published by Marcel Dekker.

ACKNOWLEDGMENTS

The authors take this occasion to express their sincere gratitude to Dr. P. A. Thrower for his constant encouragement during the preparation of the manuscript. We are indebted to Mr. N. Marshall for the translation.

REFERENCES

1. C. Kittel, *Introduction to Solid State Physics*, John Wiley, New York, 1966.

2. J. M. Ziman, *Principles of the Theory of Solids,* Cambridge University Press, London, 1971.

3. W. Mercouroff, *Aspects modernes de la physique des solides,* Masson, Paris, 1969.

4. N. F. Mott and H. Jones, *The Theory of the Properties of Metals and Alloys,* Dover, New York, 1958.

5. J. Callaway, *Quantum Theory of the Solid State*, Academic Press, New York, 1974.

6. J. Callaway, *Energy Band Theory,* Academic Press, New York, 1964.

7. I. L. Spain, in *Chemistry and Physics of Carbon,* Vol. 8, Marcel Dekker, New York, 1972.

8. D. F. Johnston, *Proc. R. Soc. Lond. A227,* 349 (1955).

9. J. D. Bernal, *Proc. R. Soc. A106,* 749 (1924).

10. C. Maughin, *Bull. Soc. Fr. Min. 49,* 32 (1926).

11. A. Cotton, *Chemical Applications of Group Theory,* Wiley, New York, 1970.

12. D. L. Schonland, *Molecular Symmetry,* Van Nostrand-Reinhold, New York, 1971.

13. F. Bassani and G. Pastori Parravicini, *Nuovo Cimento 508*(1), 95 (1967).

14. J. F. Cornwell, *Group Theory and Electronic Energy Bands in Solids,* North Holland, Amsterdam, 1969.

15. J. C. Slater, *Phys. Rev. 36,* 57 (1930).

16. G. Watelle, P. Granger, and G. Bertrand, *Exercices et problèmes de chimie générale,* Armand Colin, Paris, 1972.

17. P. Barchewitz, *Spectroscopie atomique et moléculaire,* Masson, Paris, 1971.

18. C. Cohen-Tannoudji, B. Diu, and F. Laboë, *Mécanique quantique,* Hermann, Paris, 1973.

19. L. J. Schiff, *Quantum Mechanics,* McGraw-Hill, New York, 1968.

20. F. A. Berezin, *The Methods of Second Quantization,* Academic Press, New York, 1966.

21. C. Ytzykson and J. B. Zuber, *Quantum Field Theory,* McGraw-Hill, New York, 1950.

22. W. Mercouroff, *La surface de Fermi des métaux,* Masson, Paris, 1967.

23. H. Jones, *The Theory of Brillouin Zones and Electronic States in Crystals,* North Holland, Amsterdam, 1975.

24. S. Mrozowski, *Phys. Rev. 92,* 1320 (1953).

25. C. A. Coulson, *Hückel Theory for Organic Chemists,* Academic Press, New York, 1978.

26. M. Bradburn, C. A. Coulson, and G. S. Rushbrooke, *Proc. R. Soc., Edinburgh* 336 (1946).

27. C. A. Coulson, *Proc. R. Soc., Edinburgh* 210 (1941).

28. C. A. Coulson and R. Taylor, *Proc. Phys. Soc.* A65, 815 (1952).

29. J. C. Slonczewski and P. R. Weiss, *Phys. Rev. 109,* 272 (1958).

30. J. W. McClure, *Phys. Rev. 119,* 606 (1960).

31. M. S. Dresselhaus and J. G. Mavroïdes, *IBM Res. Develop. 8,* 262 (1964).

32. P. R. Schroeder, M. S. Dresselhaus, and A. Javan, *J. Phys. Chem. Solids 32,* 139 (1971).

33. J. W. McClure, *J. Phys. Chem. Solids 32,* 127 (1971).

34. P. R. Wallace, *Phys. Rev. 71,* 622 (1947).

35. S. Mrozowski, *Phys. Rev. 77,* 838 (1950); *85,* 609 (1952); *86,* 1056 (1952); *J. Chem. Phys. 21,* 492 (1953).

36. *Les Carbones par le GFEC,* Vols. I and II, Masson (see particularly Chap. VI), 1965.

37. A. Charlier, M. F. Charlier, F. Dujardin, and J. P. Decruppe, *Bull. Soc. Chim. 3,* 94 (1979).

38. O. Gonzalez and F. Dujardin, Thèses de Doctorat de 3ème Cycle, University of Metz.

39. G. F. Koster, *Solid State Phys. 5,* 210 (1957).

40. H. Goldstein, *Classical Mechanics,* Addison Wesley, New York, 1957.

41. R. Courant and D. Hilbert, *Methods of Mathematical Physics,* Interscience Publishers, New York 1953.

42. L. Pauling and E. B. Wilson, *Introduction to Quantum Mechanics,* McGraw-Hill, New York, 1935.

43. J. Bass, *Cours de mathématiques,* Masson, Paris, 1968.

44. J. M. Luttinger and W. Kohn, *Phys. Rev. 97,* 869 (1955).

45. N. Ganguli and K. L. Krishnan, *Proc. R. Soc. London A177,* 168 (1941).

46. J. E. Hove, *Phys. Rev. 100,* 645 (1955).

47. J. W. McClure, *Phys. Rev. 104,* 666 (1956); *108,* 612 (1957).

48. R. E. Peierls, *J. Phys. 80,* 763 (1933).

49. A. H. Wilson, *Proc. Cambridge Phil. Soc. 49,* 292 (1953).

50. E. N. Adams, *Phys. Rev. 89,* 633 (1953).

51. L. Landau and E. Lifchitz, *Mécanique quantique,* Editions Mir, 1966.

52. A. Messiah, *Mécanique quantique,* Dunod, Paris, 1969.

53. J. M. Luttinger, *Phys. Rev. 102,* 1030 (1956).

54. M. Inoue, *J. Phys. Soc. Japan 13,* 382 (1958); *17,* 808 (1962).

55. R. R. Haering and P. R. Wallace, *J. Phys. Chem. Solids 3,* 253 (1957).

56. J. M. Ziman, *Elements of Advanced Quantum Theory,* Cambridge University Press, London, 1969.

57. R. R. Haering, Thesis, McGill University, The electric and magnetic properties of graphite (1957).

58. C. Chahine and P. Devaux, *Thermodynamique statistique,* Dunod, Paris, 1970.

59. P. Robert, *Traité d'électricité,* Vol. II, *Matériaux de l'électrotechnique,* Editions Georgi, 1979.

60. Carter and Bate, Eds., *The Physics of Semi-Metals and Narrow-Gap Semiconductors,* Pergamon Press, Oxford, 1971: J. W. McClure, pp. 127-137; P. R. Schroeder, M. S. Dresselhaus, and A. Javan, pp. 139-148; A. D. Boardman and G. Graham, pp. 149-164.

61. K. Shugihara, *J. Phys. Soc. Japan 18,* 322 (1963).

62. K. Shugihara and S. Ono, *J. Phys. Soc. Japan, 21,* 631 (1966).

63. A. H. Kahn and H. P. R. Frederikse, *Solid State Phys. 9,* 258 (1959).

64. P. Nozières, *Phys. Rev. 109,* 1510 (1958).

3

Interactions of Carbons, Cokes, and Graphites with Potassium and Sodium

HARRY MARSH, NEIL MURDIE,[*] IAN A. S. EDWARDS

Northern Carbon Research Laboratories
University of Newcastle upon Tyne
Newcastle upon Tyne, United Kingdom

HANNS-PETER BOEHM

Institut fur Anorganische Chemie, Universitat München
München, Federal Republic of Germany

[*]Present affiliation: Materials Technology Center, Southern Illinois University, Carbondale, Illinois

I. INTRODUCTION

This review summarizes current understanding of the mechanisms of
intercalation of potassium and sodium with graphitic, graphitizable,
and nongraphitizable carbons. The objective is to facilitate research
into industrial aspects of the interactions of alkali metals with
metallurgical cokes and anodes and cathodes employed in aluminum
production. The role of alkali materials in the blast furnace is
outlined, and a summary is provided of current understanding of
intercalation and staging in graphite intercalation compounds.

Potassium appears to be able to enter into all carbon forms
except graphitizable carbon of low heat treatment temperature.
Sodium has been reported to form intercalates with graphite and
appears to enter into both the graphitizable and nongraphitizable
carbons much more easily. For single-crystal graphite, resultant
potassium intercalates exhibit maximum structural ordering, the
staging of the intercalation compounds being well defined. With
other forms of carbons, potassium probably enters via the same electron transfer reaction as that for graphite. However, because these
carbons are less structured than graphite, precise identification of
intercalation compounds and their staging cannot be made. Carbons
are found to swell following interaction with potassium. The presence of porosity within nongraphitizable carbons facilitates entry
of potassium vapor. Swelling may cause internal stresses and

comminution of pieces of carbon. The potassium compounds are
unstable in atmospheres of CO_2, O_2, and H_2O. Potassium can be
retained in graphitizable and nongraphitizable carbons to higher
temperatures (~1000°C) than in graphitic carbon (~600°C).

Interest in the preparation and properties of intercalation
compounds of graphitic and graphitizable carbons with potassium and
sodium as intercalates has grown, and fundamental knowledge
of the properties of these materials has increased significantly.
This renewed interest has been initiated by current research into the
effects of alkali (particularly those containing potassium) causing
degradation of metallurgical coke in the blast furnace and degradation of electrodes used in aluminum production. The literature
describing intercalation is extensive but has not previously been
critically assessed. Therefore, it is appropriate to review this
subject and its relevance to industrial problems.

Graphite intercalation compounds are formed by the insertion of
atomic or molecular layers of a different chemical species between
layers in a graphitic host material. An important property of graphite intercalation compounds is "staging," which is the phenomenon of
intercalate layers periodically arranged in a matrix of host layers.
Graphite intercalation compounds are therefore characterized by a
stage index n denoting the number of host layers between adjacent
intercalated layers, as illustrated in Fig. 1. This staging phenomenon is diagnostic for graphitic and graphitizable carbons alike even
in those carbons with dilute intercalation concentrations (n ~ 10).

More than 200 compounds can be intercalated into graphitic
carbons. They are classified according to whether they donate electrons to or accept electrons from the carbon layer forming donor or
acceptor compounds. Potassium is one of the most common and widely
studied of the donor elements; others are rubidium, cesium, alkaline
earth metals, certain rare earth metals, such as europium, and metal
alloys of these. Ternary donor intercalation compounds have also
been prepared using alkali metals with hydrogen or with polar molecules, such as ammonia and tetrahydrofuran.

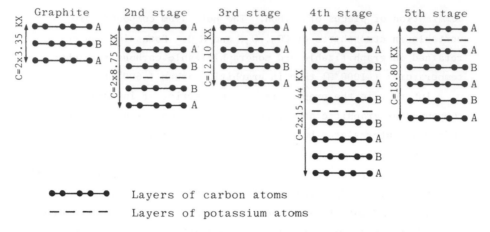

Layers of carbon atoms

Layers of potassium atoms

FIG. 1   Arrangements of carbon hexagon networks and potassium ions (atoms) in graphite intercalation compounds showing staging; viewed perpendicular to the c axis.   *Note*: 1KX = 100 pm.   (From Ref. 3.)

Nomenclature for Solid Carbons

In preparing a review of interaction of potassium and sodium with "carbons," it is advisable to use the nomenclature recently created to describe the forms of carbon that exist (see *Carbon 20,* 445-449, 1982).

Solid carbon (usually referred to in texts as "carbon") covers all natural and synthetic substances consisting mainly of atoms of the element carbon and with the structure of graphite or at least with two-dimensionally ordered layers of carbon.

A char is defined (*Carbon 21,* 517-519, 1983), as a carbonization product of a natural or synthetic organic material that has not passed through a fluid stage during carbonization.

A coke is a highly carbonaceous product of pyrolysis of organic material at least parts of which have passed through a liquid or liquid crystalline state during the carbonization process and that consists of nongraphitic carbon.

Nongraphitic carbons are all varieties of substances consisting mainly of the element carbon with two-dimensional long-range order of the carbon atoms in planar hexagonal networks but without any measurable crystallographic order in the third direction (c direction), apart from more or less parallel stacking.

Graphitic carbons are all varieties consisting of the element carbon in the allotropic form of graphite irrespective of the presence of structural defects and characterized by distinguishable three-dimensional order, recognized by at least some modulation of the (hk) x-ray reflections.

Nongraphitizable carbon is a nongraphitic carbon that cannot be transformed into graphitic carbon solely by high-temperature treatment up to 3300 K under atmospheric or lower pressure.

Graphitizable carbon is a nongraphitic carbon that, upon graphitization heat treatment, converts into graphitic carbon.

Anisotropic carbon and isotropic carbon are also used to describe carbon forms in terms of the properties of polished surfaces when viewed by polarized light optical microscopy. The anisotropic carbons are graphitic or graphitizable, their polished surfaces exhibiting optical activity. The polycrystalline nature of graphitizable carbons can be distinguished and is referred to as optical texture. The isotropic carbons are the nongraphitizable carbons, their polished surfaces exhibiting no optical activity.

II. CARBON-ALKALI METAL SYSTEMS

A. Intercalation of Alkali Metals with Graphites

The reactions of carbons with potassium were first reported in 1926 by Fredenhagen and Cadenbach [1]. In their studies they reacted potassium vapor with a natural flake graphite in a two-zone glass apparatus. The apparatus was first evacuated for 12 hr. The potassium metal was then heated to temperature $T_i = 250°C$ in an ampoule separated by a constriction from a second ampoule containing the flake graphite at a higher temperature (T_c) (Fig. 2). The authors

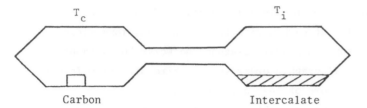

Carbon   Intercalate

FIG. 2 Schematic diagram of the two-zone vapor transport method, where T_c and T_i are the temperatures of the carbon and intercalate respectively.

observed that when the potassium vapor was in contact with the flake graphite, there was a change in color of the graphite. Although they were unable to obtain the exact composition of the compound formed, they made the following important observations.

Fredenhagen and Cadenbach were the first to show that the extent of K, Rb, and Cs alkali metal intercalation was dependent on the temperature and alkali metal vapor pressure and that the resultant compounds (stages) could be distinguished from one another in color. Such compounds were prepared by carefully distilling off in vacuum the excess of alkali metal used in the preparation. The compounds prepared at the lowest temperature had a bronze-copper red color and had the approximate composition MC_8 (M = K, Rb, Cs). On further heating to 350°C, MC_8 lost alkali metal and changed into a steel blue compound. Finally, all the alkali metal was driven off by heating to 500°C. Furthermore, Fredenhagen and Cadenbach [1] noticed that under the same experimental conditions sodium did not wet graphite although sodium reacted with soot, a relatively disordered form of carbon.

In 1932, Schleede and Wellmann [2] first showed by application of XRD (x-ray diffraction) that the products of the reaction of graphite with alkali metals K, Rb, and Cs were intercalation compounds. At this period of time little work had been done on graphitizable and nongraphitizable carbons. Using the two-zone vapor transport method, Schleede and Wellmann [2] first derived the formula KC_{16} from x-ray diffraction measurements for the blue compound.

However, further analytic and x-ray investigations by Rudorff and Schulze [3] showed conclusively that the blue compound contained less alkali metal and had the formula KC_{24}. This conclusion was supported by Herold [4], who used the two-zone vapor transport method to measure the isobars of potassium intercalation. Industrial Savoie Acheson Graphite (0.2% ash), other artificial graphites (0.55% ash) supplied by the Atomic Energy Commission, and natural Ceylon graphites (3.5% ash) were reacted with potassium. The potassium was maintained at 250°C with the graphite at temperatures in the range 250-600°C. Figure 3 shows how the composition and the stages of the intercalation compounds vary with the reaction temperature of the graphite at constant vapor pressure of potassium. The first clear

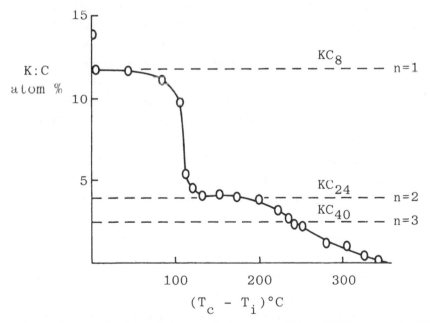

FIG. 3 The variation of intercalate composition (K/C, atom %) with increasing temperature of reaction of the graphite $(T_c - T_i)$ °C at constant vapor pressure of potassium. T_i = 250°C. (From Ref. 4.)

inflection in the composition profile occurred at KC_{24}. This product
gave the same x-ray powder pattern as the blue compound studied by
Schleede and Wellmann [2].

When studying different carbon forms, distinctions have to be
made between nongraphitic and graphitic carbon. In 1951, Franklin
[5] established, in an x-ray study, previous reports by Hofmann et al.
[6] of the existence of the two classes of nongraphitic carbon, that
is, the graphitizable and nongraphitizable carbons. The former carbons contain long-range structural order of sizes of up to ~500 μm;
the nongraphitizable carbons have crystallographic order over distances of nonometers only.

Fredenhagen and Cadenbach [1] had already observed uptake of
potassium by carbon black and by active charcoal in 1926.

In 1960, Platzer-Rideau [7] studied the occurrence of intercalation of these two forms of carbon by alkali metals. In her
studies, she used the two-zone vapor transport method as described
previously for graphitizable and nongraphitizable carbons. The
temperature of the potassium was maintained at 250°C, with a higher
reaction temperature up to 500°C maintained for the carbon.

The graphitizable carbons used were obtained by pyrolysis of
polyvinyl chloride at 1000°C. The resultant isobars indicated that
intercalation and insertion of potassium occurred in the graphitizable carbons at 500°C.

In order to study the influence of heat treatment temperature
(HTT) of the carbon (i.e., extent of graphitization), the decomposition product of polyvinyl chloride was further heat treated to 1200,
1400, 1600, 1800, and 2100°C. The variations of composition (C/K)
with the temperature of reaction ($T_c - T_i$ °C) are shown in Fig. 4
and Table 1. With increasing heat treatment temperature of the carbon, the C/K ratio for first- and second-stage compounds approached
the values for graphite of KC_8 and KC_{24}, respectively (Table 1).
Figure 4 also shows that the stability of potassium in the carbon
apparently decreases for a given reaction temperature with increasing
heat treatment temperature of the graphitizable carbon.

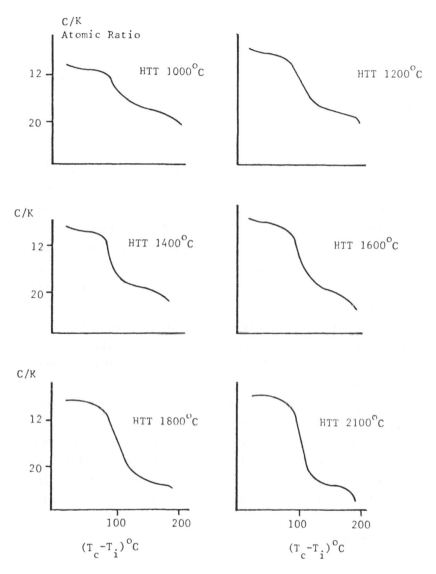

FIG. 4 The variation of composition of carbon-potassium systems with increasing temperature of reaction of the carbon $(T_c - T_i)$ °C at constant vapor pressure of potassium for polyvinyl chloride carbons of increasing HTT. $T_i = 250$°C. (From Ref. 7.)

TABLE 1   Influence of Heat Treatment Temperature of Polyvinyl
Chloride Carbons on C/K Composition (n Values in KC_n [7])

Heat treatment temperature (°C)	1000	1200	1400	1600	1800	2100	Graphite ideal
First stage (KC_8)	12.5	11.3	10.4	10.0	9.6	8.8	8
Second stage (KC_{24})	17.5	17.8	18.4	18.8	20.4	23.0	24

In her study of nongraphitizable carbons, Platzer-Rideau [7]
used a saccharose coke, HTT 1000°C. When reacted with potassium
vapor using the two-zone vapor transport method, no change in color
of the carbon was observed. The curves (isobars) obtained were
completely different from those of the graphitizable carbons (Fig.
5). There was significant uptake of potassium as KC_{10}. For a given
HTT of the carbon, the extents of potassium uptake decreased with
increasing reaction temperature ($T_c - T_i$) (Fig. 5). The carbon of
HTT 2100°C was less able to retain potassium than the carbon of HTT
1000°C. A slight change was observed in color to a blue-red, but
no swelling was reported. The author [7] concluded that no intercalation had occurred even though appreciable amounts of potassium
had reacted (Fig. 5). Potassium also reacted with glassy carbons,
causing disruption of the structure [8].

Note: The criterion of color change to indicate formation of
intercalation compounds may only be relevant for the graphitic carbons.

Platzer-Rideau [7] found that sodium did *not* intercalate in
graphitizable carbons; it was thought to be adsorbed on internal
surfaces.

These studies, overall, using potassium and sodium, gave no
clear explanation of the mechanism of entry of the alkali metal into
the nongraphitizable carbons (Fig. 5). The outstanding industrial
problem that led to these studies, namely, disruption of carbon
electrodes in contact with sodium during the processing of aluminum,
was not explained. The further pursuit of these aspects was the
motivation behind the research of Berger et al. [9] and other workers
[10-13] into interactions of carbons with alkali metals.

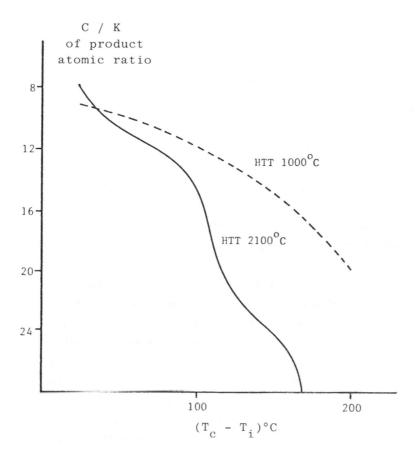

FIG. 5 The variation of composition of carbon-potassium systems with increasing temperature of reaction of the carbon $(T_c - T_i)$ °C at constant vapor pressure of potassium for saccharose carbons, HTT 1000°C and 2100°C. T_i = 250°C. (From Ref. 7.)

Berger et al. [9] proposed that the term "intercalation" be used when a reacted species is inserted between graphitic layers that have not been moved aside prior to reaction. Intercalation is thus characterized by a finite deformation of the carbon, the process being reversible. On the other hand, carbons that are porous and nongraphitizable may be able to assimilate material, such as potassium, within the substance of the carbon, not by a process of intercalation but by adsorption into the porosity. It is necessary to distinguish between these two processes.

B. Interaction of Potassium with Graphitizable Carbons

Although the properties of potassium-graphite are now relatively well established and understood, the formation and properties of compounds formed from potassium with nongraphitic carbons are less well documented.

It has been reported [9,12,14] that nongraphitic graphitizable carbons, such as petroleum cokes and carbon fibers (PAN fibers and pitch fibers), form intercalation compounds with potassium.

Caston and Herold [8], using x-ray and thermogravimetric analysis methods, studied potassium intercalation in graphitizable petroleum cokes of HTT 1250, 1600, 2000, and 2400°C. The potassium was kept at 200°C, with the carbon maintained at a higher temperature than the potassium up to a maximum temperature of 800°C. Thermogravimetric analysis indicated that all these carbons interacted with potassium, and this resulted in swelling. The volume increase was found from the dimensions of the carbon piece before and after reaction. For graphite and graphitizable carbons, a volume expansion of 55% was observed, a value close to the theoretical value for KC_8 (61%). Thermogravimetric analysis indicated that the saturated compounds (prepared at p/p^o close to 1) were richer in potassium than KC_8. Their x-ray diffraction patterns were characteristic of a first-stage compound, indicating that intercalation was accompanied by only limited adsorption. They found that compounds formed with graphitizable carbons were similar to the graphite compounds. The modification of the x-ray diffraction patterns with increased HTT of the carbon enabled them to index the lines and the diffuse maxima observed.

For the intercalated graphite KC_8 and the intercalated 2000 and 2400°C cokes, a set of diffraction lines was observed between the (001) and (002) lines. These lines were not resolved in the 1250 and 1600°C cokes, but in their place a broadened and asymmetric maximum occurred.

The x-ray diffraction patterns of compounds lower in potassium content indicated a mixture of the first and second stages. The pattern of a second-stage compound did not show the $(11)_K$ band corresponding to the metal sublattice as with potassium graphite of stage >2. The intercalate structure is thought to be of a disordered lattice gas nature. The thickness of the potassium layer was identical to the thickness measured for the first-stage compound. Thus, the compounds prepared from the graphitizable carbons are similar to the graphite compounds.

Huhn et al. [15] treated cokes with potassium metal at elevated temperatures in an inert atmosphere. Considerable quantities of potassium were bound by the carbons up to 900 K, corresponding to a stoichiometry of $KC_{5.5}$. This material gave off potassium at temperatures as high as 827°C (1100 K).

Terai and Takahashi [13] further established that the interlayer spacing and distribution of potassium atoms depended on the structure and the chemical composition of the initial carbon materials. Using an adaption of the two-zone vapor transport technique, they reacted petroleum cokes of HTT 800, 1000, 1500, and 2300°C and, for comparison, a natural graphite with potassium at 300°C for 48 hr. The mass ratio of the carbon to potassium was controlled to yield the KC_x compound of the desired composition. The KC_x composition was measured by gravimetric and chemical analysis. The KC_x compounds were prepared under such conditions that the mole ratio of the carbon-potassium mixture in the reaction tube varied from 3.5 to 18. The compositions of resulting compounds [(C/K) product] are plotted against the C/K ratio of the reaction mixture in Fig. 6. Figure 6 shows three plateaus: the first and third plateaus indicate the KC_8 phase (stage 1) and the KC_{24} phase (stage 2), respectively. These two plateaus are also found using carbons of higher HTT (e.g., 2300°C), that is, the graphitic carbons (see Fig. 3). For the coke heat treated at 1500°C, however, another phase between KC_8 and KC_{24} is suggested by the second plateau, the composition of this phase corresponding to KC_{12}-KC_{14}.

With the low-HTT carbons, specific intercalation compounds with stoichoimetric compositions were not found. Although $KC_{8.2}$

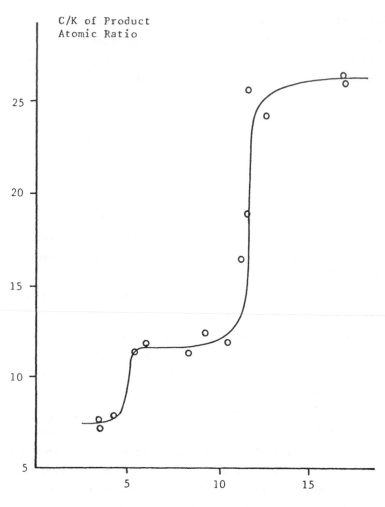

FIG. 6 The variation of product composition (C/K) with reactant mixture composition (C/K) for petroleum coke (HTT 1500°C) with potassium at 300°C for 48 hr. (From Ref. 13.)

(HTT 1000°C) and $KC_{14.3}$ (HTT 1000°C) were identified, the plateaus, as shown in Fig. 6, were not apparent in comparable plots. No potassium intercalation compounds were formed for HTT 800°C cokes. X-ray analysis of the carbons of HTT 1000 and 1500°C gave broad diffraction lines, suggesting only short-range order in the interplanar spacings.

Kwizera et al. [14] investigated the properties of graphitic carbon fibers as modified by intercalation. The fibers were heated at 300°C in evacuated tubes containing the vapors of K, Rb, or Cs using the two-zone temperature method. Results of Raman spectroscopy and x-ray diffraction were reported for both pristine and intercalated PAN-based and pitch-based fibers. X-ray analysis indicated the pitch-based fibers to be more highly ordered than the PAN-based fibers. Raman results confirmed this and were consistent with evidence from transmission electron microscopy that the microstructure of both types of pristine fibers consist of "crystalline and disordered graphite regions" (i.e., graphitic and nongraphitic regions).

X-ray results gave direct evidence for staging in alkali metal donor intercalated fibers through the appearance of "superlattice reflections" corresponding to the packing in the alkali layers but suggested that the staging was mixed in the fiber bundles as shown by Raman spectroscopy. However, there is evidence for single-stage regions occurring in the optical skin depth only. Comparison of the two methods of analysis indicated the skin region to be better staged than the core, a result that is also consistent with the higher degree of ordering in the skin relative to the core for both types of fiber.

C. Interaction of Potassium with Nongraphitizable Carbons

Caston and Herold [8] investigated the possibilities of intercalating various nongraphitizable carbons, including vitreous carbons, saccharose carbons, and calcined material obtained from a Vietnamese anthracite. These cokes have a considerable porosity and a large surface area. Their properties are summarized in Table 2.

Berger et al. [9] proposed that because carbons are known to have an ultramicroporosity, it was illusory to distinguish between the condensation of a metal in this porosity (where the walls are to a great extent basal plane layers) and intercalation. In addition to condensation in porosity, the small-volume elements of short-range order in nongraphitizable carbons, as described by Ban et al. [16], may be a suitable environment for the potassium to intercalate.

TABLE 2   Properties of Nongraphitizable Carbons Studied by
Caston et al. [8][a]

Carbon	d_{He} (g cm^{-3})	d (002) (pm)	Closed porosity (cm^3 cm^{-3})	P_x (cm^3 cm^{-3})	S_x (m^2 g^{-1})
Vitreous carbon, 1000°C	1.47	364	0.285	0.176	540
Saccharose coke					
1000°C	2.03	367	0.034	0.175	460
1500°C	1.49	363	0.034	0.175	690
2000°C	1.49	354	0.331	0.245	580
2500°C	1.46	351	0.340	—	—
Anthracite coke, 1650°C	1.75	354	0.198	0.123	330
Vitreous carbon, 2500°C	1.44	348	0.337	0.288	480

[a]P_x = porosity (small-angle x-ray scattering). S_x = surface area =
$4P_x(1 - P_x)l_p^{-1}v$; l_p = mean length of pore; V = pore volume.

Thermogravimetric analysis indicated that in all cases relatively large quantities of potassium in these carbons were stable to
high outgassing temperatures [9].   However, the volume of liquid
metal "adsorbed" by the carbon was greater than the pore volume of
the carbon, indicative of swelling or imbibition of the carbon.   All
the nongraphitizable carbons in that study formed the first-stage
compound KC_{10}, accompanied by limited adsorption, for all heat treatment temperatures of carbon preparation.

X-ray analysis showed the appearance of a new band (10)$_K$ between
the (001) and (002) lines.   The (001) line is weak and disappears
into the (10)$_K$ band at the lower heat treatment temperature.   Berger
et al. [9] postulated, therefore, that these nongraphitizable carbons
are able to form some type of heterogeneous intercalated material
(not necessarily a compound).

D.   Influence of Reaction Temperature on the
     Formation of Potassium Intercalation Compounds

In most intercalation studies of potassium, the preparation of a
required compound has taken place isothermally at temperatures

below 600°C. This is mainly because at temperatures in excess of
600°C, potassium reacts with the Pyrex ampoules. However, it has
been suggested by Hennig [17] that intercalation can occur at temperatures higher than 600°C.

It has been shown unambiguously that potassium carbonate can
be reduced to metallic potassium by carbon at elevated temperatures
[18-22]:

$$K_2CO_3 + 2C \rightarrow 2K + 3CO$$

Formation of intercalation compounds in K_2CO_3-catalyzed gasification
reactions cannot be excluded a priori. The thermal stability of a
given stage depends on the ratio p/p^o of the potassium vapor pressure
to its saturation pressure at the reaction temperature. The thermodynamic activity of potassium is considerably reduced by intercalation, the more dilute in potassium is the compound. At low values
of p/p^o, potassium may be taken up by the graphite or carbon structure
without forming discrete new phases (stages). The potassium uptake
will be facilitated by the presence of lattice defects and especially
by the low crystalline order of carbons compared with graphite. Such
compounds may be considered dilute solutions of potassium in carbon
or graphite [18]. Very little is known about such dilute systems.
The electrons transferred to the carbon layers may influence their
chemical reactivity to some extent.

Yokoyama et al. [23] heated an undefined "amorphous" carbon of
78 m^2 g^{-1} surface area in vacuo with potassium carbonate to 650°C and
studied the reaction product using the technique of x-ray photoelectron spectroscopy (XPS). After this treatment, the K 2p and O 1s
peaks had diminished in intensity considerably, but they could be
restored to high intensity by a pressure of about 50 Pa of carbon
dioxide at 650°C or under a low pressure of oxygen at room temperature. On reheating in vacuo, the XPS signals became very weak again.
The authors concluded that potassium metal was formed by reduction of
the carbonate and that the potassium diffused into the bulk of the
carbon. As with intercalated potassium, it was reactive toward oxidizing gases (Sec. II.F).

Kapteijn et al. [24] carried out an x-ray diffraction study on
the effect of alkali metals from alkali carbonates on graphite, on
an activated carbon, and on a semianthracite coal at 900-1200 K
(627-927°C) in an inert gas atmosphere. Reaction of a dry mixture
of activated carbon (Norit RX extra), graphite, or a coal (semianthracite and VI-DE, SBN) with alkali carbonate in a molar ratio of
46:1 was carried out under flowing nitrogen. The system was heated
at 30 K min^{-1} to 1100 K (827°C), followed by isothermal heating for
30 min. The samples were cooled and transferred to Debye-Scherrer
capilliaries under nitrogen.

Model intercalation compounds (KC_8, KC_{24}, and KC_{36}) were also
prepared from graphite and metallic potassium using the method
described by Podall et al. [25]. Graphite was reacted for 40 min
with potassium (chemically pure reagent) under stirring at 275 ± 20°C.
The effect of heat treatment to 827°C on the activated carbon without
carbonate or metal was found to be insignificant. Heating with potassium carbonate rendered the carbons and coals pyrophoric. The authors
concluded from the x-ray diffraction pattern that structures similar
to a stage 3 compound (KC_{36}) had formed to some extent. However,
Fig. 2 of their paper shows sharp x-ray reflections superposed on
the broad diffraction lines of the parent activated carbon. It seems
very unlikely that well-ordered structures with large crystallite
dimensions should form upon intercalation. The sharp reflections
were probably owing to products of the reaction of potassium with
contaminants after heating. The pyrophoric nature could have been
caused by ill-defined dilute intercalation compounds or by finely
dispersed metallic potassium adsorbed on the porous carbon. Heat
treatment of the coal in the absence of potassium to 1100 K yielded
more pronounced x-ray reflections, indicating an ordering within the
structure of the resultant coke. Disappearance of these reflections
for the coal occurred after reaction with potassium. This was accompanied by a disintegration of the char particles during preparation.

The results of Kapteijn et al. [24] showed that potassium can
break up the char structure into highly disordered carbon. No intercalation structures were formed by heat treatment of K_2CO_3-graphite

samples, as confirmed by Ferguson et al. [26]. Kapteijn et al. [24] found that alkali metals were responsible for the breakup of the coal structure, this becoming more pronounced with increased concentration of metal in the order K < Rb < Cs. In contrast, Li and Na were unable to change the structure. They concluded that the breakup was due to the formation of intercalation compounds. The supporting evidence for this statement is, however, not strong.

In contrast to char and carbon, graphite was apparently unable to reduce potassium carbonate to potassium metal. Kapteijn et al. [24] explained this phenomenon by the presence in chars or activated carbons of polar surface groups that enhance the decomposition of the carbonate and the reduction to the metallic state. Wigmans et al. [27] also concluded that these polar surface groups may be responsible for the higher stability of the potassium metal in the char compared with the graphite because it is known that alkali metal-graphite intercalation compounds decompose upon heat treatment by evaporation of the metal [2]. The graphite surface is itself relatively unreactive to the formation of metallic potassium and intercalates.

Rao et al. [18], studying the catalysis of the Boudouard reaction by alkali metal compounds, considered further the suggestion of Wen [28] that intercalation compounds take part in the reaction mechanism according to the following scheme:

$$M_2CO_{3(s,l)} + 2C = 2M_{(g)} + 3CO_{(g)}$$

$$2M_{(g)} + 2nC_{(s)} = 2MC_{n(s)}$$

$$2MC_{n(s)} + CO_{2(g)} = (2nC) \cdot M_2O_{(s)} + CO_{(g)}$$

$$(2nC) \cdot M_2O_{(s)} + CO_{2(g)} = 2nC_{(s)} + M_2CO_{3(s,l)}$$

This scheme of reaction stages assumes that the intercalation compounds of the type MC_n are stable at the reaction temperatures and atmospheres normally used in the study of the catalysis of the Boudouard reaction. These atmospheres, like those in the blast furnace, are rich in oxygen-containing compounds, unlike the inert atmospheric conditions of the established intercalation studies.

In assessing if intercalation compounds could be intermediates, Rao et al. [18] noted the lack of reliable thermochemical data pertaining to intercalation compounds at these temperatures (527-1127°C). The data of Berger et al. [9] and Aronson et al. [29] do not extent to these temperatures.

Using standard available thermochemical data, Rao et al. [18] calculated values of n (in MC_n) for compounds in equilibrium with a gas phase containing carbon dioxide. They concluded that, in the presence of carbon dioxide, intercalation compounds are not predisposed to form until temperatures of 1173°C or higher are considered. However, in an environment free of carbon dioxide, intercalation compound formation seems to be possible [21].

Moulijn et al. [30] presented x-ray evidence that potassium-graphite intercalation compounds are not stable above 397-427°C in an atmosphere of flowing helium.

In a review of the mechanism of the catalysis by alkali carbonates of the gasification of coal chars, Wood and Sancier [31] concluded that intercalated potassium is not significantly involved in the catalytic step because intercalation compounds are not stable under reaction conditions. Wood et al. [31,32] favor a reaction mechanism involving a molten alkali suboxide. The authors reported that the electrical conductivity of a coal char impregnated with potassium carbonate increased by more than two orders of magnitude on heating in helium to 727°C. After this treatment, the conductivity increased still further on cooling to 27°C, reaching a value of about 0.3 Ω^{-1} cm^{-1}. The negative temperature coefficient of electrical conductivity was confirmed on repeated heating-cooling cycles. The conductivity decreased upon addition of carbon dioxide. The authors attributed the electrical conductivity to a layer of molten potassium salt. This seems not very plausible at 27°C, and the observed behavior seems to be indicative of a dilute intercalation compound or of another metallic constituent. Cesium suboxides are metallic, but analogous potassium compounds are not known.

Hawkins et al. [22] studied the reaction of potassium and sodium vapors in a gaseous environment of nitrogen or carbon monoxide with

petroleum coke, metallurgical cokes, anthracite carbons, and graphites
in a range of reaction temperatures from 572 to 1200°C. Their principal conclusions were as follows.

1. The reaction occurred with the carbon material, not with
    mineral matter contained within the carbon.
2. Disintegration of all forms of carbon became severe at the
    lower temperatures.
3. Graphitic carbons were more resistant to attack by potassium
    than nongraphitic carbons.
4. The process of attack was probably through intercalation at
    low temperatures, although at high temperatures, charge-
    transfer reactions might be possible (*Note*: intercalation
    is a charge-transfer reaction, and it is not reasonable to
    so separate two mechanisms.)
5. Concentration gradients of potassium within a carbon may
    be responsible for some disruption.
6. Calcined anthracites have a high resistance to attack.

Reaction with potassium in the presence of carbon monoxide (carbon
dioxide was not studied) apparently was indistinguishable from reaction in argon.

E. Rates of Intercalation

Hooley et al. [33], in 1965, investigated the kinetics of intercalation and suggested that the process of intercalation created tensive
forces between the outer surface region and the inner core region of
graphite pieces. If the tensive forces were small, then intercalation proceeded to the inner core.

   A strain energy was introduced owing to the separation of two
graphite layers by insertion of the intercalate. Intercalation did
not begin unless the vapor pressure of the reagent was greater than
the threshold pressure p_t. It is believed that a threshold pressure
was required to unpin lattice dislocations and to relieve lattice
strain, because the intercalation process required motion of dislocations [34,35]. The threshold pressure was dependent [33] on

1. The substance intercalated

2. The reaction temperature

3. The structure within the graphite material

It has been found that the order of threshold pressure is lower for single-crystal flake graphite and increases for highly orientated pyrolytic graphite and carbon fibers and that in all cases it is lowered for thin host samples.

Thermodynamic considerations have been related to the transition from one intercalation stage to another, measuring changes in free energy, enthalpy, and entropy associated with stage formation. Generally, energy is absorbed upon stage transition $n \rightarrow n - 1$ and the entropy decreases. Using the solid-state electrochemical cell method on graphite-K systems, Aronson et al. [29] obtained values for enthalpy and entropy changes associated with stage transformations (Table 3).

Kobayashi and Omari [36] investigated the rate of potassium uptake in metallurgical coke from nitrogen gas-potassium vapor mixtures. Their experimental results for rates of potassium uptake from nitrogen-potassium mixtures with 10^{-4}-10^{-3} partial pressure of potassium at 900-1100°C showed that the rate of uptake depended on the

TABLE 3 Thermodynamic Properties of Graphite-Potassium Compounds

Reaction	$-H°$ (kJ mol^{-1})	$-S°$ (J K^{-1} mol^{-1})
$1/3C_{24}M_{(s)} + M_{(s)} \rightleftharpoons C_8M_{(s)}$	27.4	25.7
$4C_{10}M_{(s)} + M_{(g)} \rightleftharpoons 5C_8M_{(s)}$	38	44
$5/7C_{24}M_{(s)} + M_{(g)} \rightleftharpoons 12/7C_{10}M_{(s)}$	24	24
$2C_{36}M_{(s)} + M_{(g)} \rightleftharpoons 3C_{24}M_{(s)}$	27.8	20.6
$3C_{48}M_{(s)} + M_{(g)} \rightleftharpoons 4C_{36}M_{(s)}$	30	20.7

Source: From Ref. 29.

size of coke particles and the partial pressure of the potassium, with a maximum rate at a temperature of 1000°C.

F. Stability of Potassium Intercalation Compounds

Potassium-graphite intercalation compounds are known to be unstable and readily decompose upon exposure to air or water. In 1926, Fredenhagen and Cadenbach [1] reported that potassium-graphite intercalation compounds (KC_n) reacted violently with oxygen, air, and water. Later, Daumas and Herold [37] investigated the chemical properties of KC_n with O_2, N_2O, NO, NO_2, SO_2, CO_2, and CO. From x-ray analysis they found that oxidation by oxygen, from room temperature to about 120°C, produced α-KO_2 and finally graphite, with KC_{24}, K_2O, and K_2O_2 formed during intermediate stages.

Recently, Underhill et al. [38] reported that the stage 5 compound of KC_n obtained by the decomposition of stage 4 compounds was stable in air.

Hennig [17] reported that most of the reactant within graphites is driven off at temperatures not much higher than the temperatures of formation. In potassium, rubidium, and cesium compounds, the alkali metal is desorbed as soon as the vapor pressure of the intercalated metal is reduced; that is, K, Rb, and Cs do not form residue compounds. Some alkali metal can be retained at the periphery of the carbon particles.

Earlier, Ubbelohde and Lewis [10] stated that alkali metal-graphite compounds were quite stable in vacuo at ordinary temperatures, although (as follows from their method of preparation) they lose alkali metal at higher temperatures as vapor. In air, they readily oxidize and even ignite spontaneously if free access of oxygen is permitted to the powders (they are pyrophoric).

Akuzawa et al. [39] studied the stability of KC_n under an oxygen atmosphere at temperatures from 0 to 60°C. Dry oxygen gas was introduced to the KC_n sample and held at constant temperature, and the pressure change of the oxygen was measured continuously. They found that KC_{24} was more reactive than KC_8. The rate of oxygen consumption by KC_8 or KC_{24} was found to follow a parabolic law with respect to time. X-ray diffraction analysis indicated both KC_8 and KC_{24} compounds were transformed to higher stage compounds by oxidation.

The authors suggest that the oxidation rate is determined by rates of diffusion of the potassium atoms from the intercalation sites to the surface, where they combine with oxygen to form KO_2.

The effect of particle size on reactivity of oxygen to KC_n was also examined. It was concluded that reactivity was proportional to the edge area of the intercalate. The studies of Akuzawa et al. [39] on stabilities in oxygen relate to the discussions of Rao et al. [18] on stabilities in carbon dioxide and those of Schlogl and Boehm [40] on the reaction of KC_8 with water. Less potassium hydroxide and hydrogen were formed in the reaction with water than expected for the reaction [41,42]

$$KC_8 + H_2O \rightarrow 8C + KOH + \frac{1}{2}H_2$$

The discrepancy was the greater the larger the particle size of the parent graphite. It has been shown [40] that potassium, potassium compounds, and hydrogen are trapped within the reaction product. X-ray diffraction showed additional reflections near graphite (002), indicating that ill-defined, high-stage potassium compounds were contained in the reaction product [26,40].

G. Interaction of Sodium with Graphitizable Carbons

Sodium, being less volatile than potassium and able to react with Pyrex glass, requires a two-zone apparatus made of refractory steel for the preparation of sodium-carbon compounds [43]. This apparatus was sealed using an alkali metal seal.

Berger et al. [9] state that "saturated" compounds were prepared at $T_{Na} = 580°C$ [$p_{Na} = 2.36$ kPa (18 torr)] and $T_c = 600°C$ ($p/p^0 = 0.75$). The metal content of the reaction products determined from sorption curves was checked by measuring weight increases after reaction and by acidimetric titration of the alkali metal. In the latter

TABLE 4   Intercalation of Sodium in Graphitizable Carbons

Carbon	C/Na (from thermogravimetric measurements)	Stage (assumed)	Nominal composition
Anthracene coke, 650°C	7.2	1	NaC_8
Petroleum coke, 1250°C	20	2	NaC_{24}
Petroleum coke, 1600°C	46	5	NaC_{60}
Petroleum coke, 2000°C	72	8	NaC_{96}

Source: From Ref. 9.

case, to take account of all the metal it was necessary to burn the product and then titrate the carbonate with H_2SO_4.

Berger et al. [9] found that sodium uptake for graphitizable carbons as measured by thermogravimetry decreased rapidly as the heat treatment temperature increased (see Table 4). The isotherms were monotonic and hysteresis was almost negligible, suggesting that capillary condensation did not occur. Graphitizable carbons swell on contact with sodium, the apparent dilation being 4% for a 1250°C HTT petroleum coke. The sodium-carbon compounds were of the same color and appearance as the carbons themselves. Homogeneous pieces of carbon were unchanged after several sorption-desorption cycles without degradation, whereas the agglomerated bricks of carbon disintegrated.

X-ray analyses indicated a displacement of the (002) line, with the increase in the c-axis parameter at a constant a-axis parameter corresponding to intercalation phenomena. Indexing of the (001) reflections led to the results listed in Table 4 and show similarities with those of Robert et al. [44], who used a sodium layer thickness of 128 pm and an atomic ratio C/Na of 10:1. However, it must be pointed out that formation of regularly stacked stages, in particular of high stage numbers, does not appear very likely in such relatively poorly organized carbons. It was not possible to observe bands corresponding to (hk) sodium, either in a superstructure (e.g., NaC_{64}) or in nonepitaxic compounds, as found in the coke-potassium compounds. Robert et al. [44] concluded that the sodium layers are not structured.

Sodium uptake in graphitizable carbons is attributed to intercalation. Adsorption and capillary condensation could account for only an insignificant fraction of the sodium content because these carbons have a small microporosity.

Asher [45] and Asher and Wilson [46] had earlier reported a new intercalation compound NaC_{64} with a low stability. This compound was formed from high-purity sodium (~100 ppm K) and natural flake, artificial, and reactor-grade graphite. The reaction was carried out at 400°C for 0.5 hr.

Pflugmacher and Boehm [47] studied the reaction of carbon blacks (Table 5) with sodium. The carbon blacks were heat treated in the temperature range 900-2700°C. Each carbon black was reacted with sodium metal under vacuum at 300°C for 5-6 hr. Excess sodium was removed by distillation at 380-400°C. The reaction product was examined by chemical analysis and x-ray diffraction.

A range of compositions was obtained from NaC_{10} to NaC_{12} for furnace and channel blacks heat treated to 900°C and NaC_{10} to NaC_{20} for those heat treated to 1150°C.

With carbon blacks heat treated at temperatures higher than 1150°C, there was a substantial decrease in the sodium uptake, until at 2000°C almost no sodium was found in the reaction products of furnace blacks. There was no "intercalation" of sodium by graphitized carbon blacks (HTT 2700°C).

X-ray analysis indicated a shift in the position of the (002) reflection for all carbon blacks that contained appreciable quantities of sodium. The average interlayer spacing increased from 350 to 394 pm and decreased with increasing HTT of the carbon blacks as amounts of sodium "intercalated" decreased.

The small increase in the average interlayer spacing (44 pm) on intercalation with sodium was surprising. The authors, however, postulated that intercalation had occurred in an irregular way.

TABLE 5   Carbon Blacks Used by Pflugmacher et al.

Carbon black	Producer	Type	Surface area $(m^2 \ g^{-1})$
Corax 3	Degussa	Furnace	76
Philblack A	Phillips Chem. Co.	Furnace	44
Spheron 6	Cabot Carbon Co.	Channel	124
CK3	Degussa	Deutches Gasruss, similar to Channel	92
Thermax	Thermatomic Carbon Co.	Thermal	
Anacarbon	Hostmann & Steinberg	Acetylene	76

Source: From Ref. 47.

Also, sodium atoms in the interior of carbon blacks of low HTT fill voids between carbon layers. High-resolution transmission electron microscopy (HRTEM) has shown that spacings between carbon layers are quite variable [3,16].

The authors concluded that all carbons of low HTT with a "reasonably good parallel ordering of layers" and little cross-linking should intercalate sodium [17].

H. Interaction of Sodium with Nongraphitizable Carbons

Investigations were carried out by Berger et al. [9] of interactions of nongraphitizable carbons with sodium (Table 2). The 600°C sorption isotherms of nongraphitizable carbons were different from those for graphitizable carbons. The shape of the isotherms corresponded to a type II BET adsorption isotherm, indicating multilayer adsorption or a cooperative reaction phenomenon. Heat treatment of the carbon (1000-2500°C) had no effect on sodium uptake.

X-ray investigations indicated that sodium did not intercalate in carbons of high-HTT (2000°C) saccharose coke and 2500°C vitreous carbon but is adsorbed in ultrafine porosity. However, low and medium heat-treated carbons were intercalated. According to Berger et al. [9], sodium was intercalated in saccharose coke HTT 1500°C to form a fifth-stage compound NaC_{10}. Vitreous carbon (HTT 1000°C) formed the fourth-stage compound NaC_{12}. Saccharose coke (HTT 1000°C) formed a second- to third-stage compound. This evidence was based on the shift of the (001) reflections toward lower angles and the presence of a secondary peak at 300 pm, which, in the polyvinylidene chloride coke, was more intense than the (001) line.

Hence, Berger et al [9] postulated that nongraphitizable carbons behaved similarly to graphitizable carbons. The sodium content of the product decreased with increasing HTT of the carbon.

The observations that sodium content decreased as the HTT of the carbon was increased were discussed by Robert et al. [44] in terms of displacement of the electron Fermi level of the carbon. With graphitized pyrocarbons, the Fermi level can be lowered by doping with boron, and graphites that are doped in this way form sodium intercalation compounds. Carbons of lower HTT have lower Fermi levels.

On the other hand, apart from the analytic data of Berger et al. [9], there is no clear evidence to distinguish intercalation of sodium from adsorption of sodium vapor in existing microporosity or microporosity created by the presence of sodium (ions).

I.   Interaction of Sodium with Boron-Doped Carbons

Studies of sodium interaction on boronated pyrocarbons were carried out at 380°C by Robert et al. [44]. The metal insertion was checked on each sample by x-ray diffraction before reaction with sodium. The amount of sodium uptake was measured for the boronated carbons, as well as the amount of boron within the carbons. Figure 7 shows the variation in the quantity of sodium taken up as a function of boron substitution. A regular increase was observed for boron concentrations higher than 0.1 atomic %. The variation in the sodium content plotted in Fig. 7 initially passed through a minimum owing to an increase in the degree of graphitization of the original carbon. This is caused by the presence of boron, which acts as a graphitization catalyst [48].

The increase in sodium sorption was correlated with a reduction in the Fermi level of 0.3 eV, caused by production of holes in the valency band of graphite.

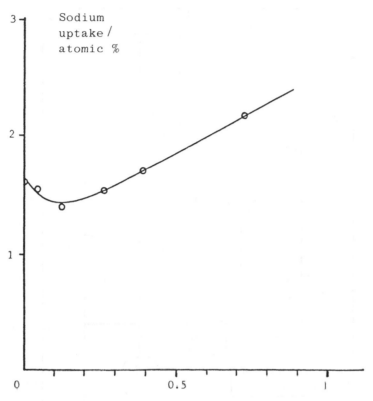

FIG. 7 Variation of sodium uptake at 380°C with extent of doping by boron for a pyrocarbon. (From Ref. 44.)

The amount of sodium inserted was found to reach a maximum for a carbon of 1000°C HTT, after which it decreased and became negligible for heat treatment temperatures >2300°C.

III. ALKALI IN THE BLAST FURNACE

The behavior of alkalis in the blast furnace has been reviewed by Davies et al. [49]. The alkali enters the furnace as complex silicates in the burden gangue and coke mineral matter (Fig. 8). The general alkali level of the materials entering the furnace is less than 0.5 wt%. These silicates in the burden decompose when they

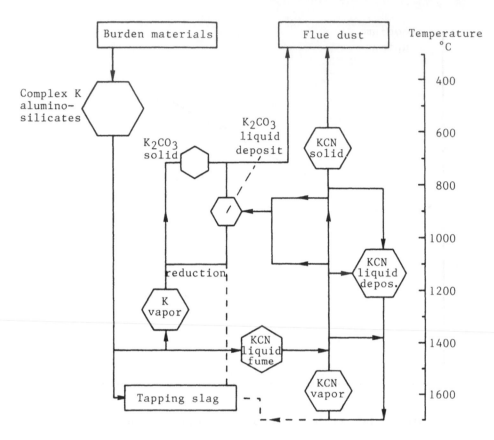

FIG. 8 Simplified mechanism for potassium recirculation within a
blast furnace. (From Ref. 49.)

reach temperatures of about 1500°C. The extent to which decomposition occurs depends upon the partial pressure of carbon monoxide and
temperature (Fig. 8). Alkali vapor produced by silicate decomposition is carried upward in the rapidly moving gas stream. Because of
the high nitrogen potential of the gas phase and the elevated temperatures, cyanides could be formed beyond the oxidizing tuyere zone:

$$2K(g) + 2C + N_2(g) = 2KCN$$

Gases leaving the hearth can be expected to contain small amounts of cyanides (KCN and NaCN), together with either free potassium and sodium vapor or their oxides. Conditions in the shaft are sufficiently oxidizing to convert cyanides and metal vapors to the more stable carbonates:

$$2K(Na)(g) + 2CO_2(g) = K_2(Na)_2CO_3 + CO(g)$$

Sodium and potassium carbonates melt at temperatures of 850 and 891°C, respectively.

Both the carbonates and the unreacted cyanides deposit on descending burden materials, coating them in dust or liquid, depending on the temperature. As the coated burden descends to the lower region of the furnace, conditions are suitable for reduction of the carbonate to metallic vapor:

$$K_2CO_3(1) + CO = 2K(g) + 2CO_2$$
$$K_2CO_3(1) + 2C(s) = 2K(1, g) + 3CO(g)$$

The potassium may return to the upper zones in the furnace to be reoxidized to the carbonate and to complete the cycle.

The recirculation and accumulation of alkalis probably occur within the temperature range 700-1200°C, above the region where the slag-forming reactions are predominant.

The "dust" obtained contains, typically, 30% alkalis and has been found at temperatures >750°C. Similarly, Willems et al. [50] indicated significant concentrations of dust containing up to 40% of alkalis in the temperature region 800-1100°C. Itaya et al. [51] reported dust loadings as high as 9.5 g m^{-3} in the 900-1200°C region of the furnace, with dust containing up to 35% potassium and 3.5% sodium. It is within the temperature region 700-1200°C that large accumulation of alkali can be expected to exert their greatest influence on refractories.

Figure 9 shows a schematic diagram of alkali distribution in the blast furnace. The deleterious effect of alkali deposits on the

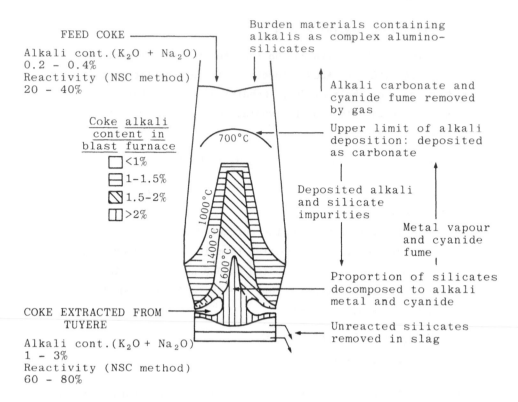

FEED COKE ─────────

Alkali cont.(K_2O + Na_2O)
0.2 - 0.4%
Reactivity (NSC method)
20 - 40%

Coke alkali
content in
blast furnace
☐ <1%
⊟ 1-1.5%
◩ 1.5-2%
▥ >2%

700°C

1000°C
1400°C
1600°C

COKE EXTRACTED FROM
TUYERE

Alkali cont.(K_2O + Na_2O)
1 - 3%
Reactivity (NSC method)
60 - 80%

Burden materials containing
alkalis as complex alumino-
silicates

Alkali carbonate and
cyanide fume removed
by gas

Upper limit of alkali
deposition: deposited
as carbonate

Deposited alkali
and silicate
impurities

Metal vapour
and cyanide
fume

Proportion of silicates
decomposed to alkali
metal and cyanide

Unreacted silicates
removed in slag

FIG. 9 Schematic diagram of alkali within the blast furnace.
(From Ref. 52.)

furnace refractories, ferruginous charges, and general operation of
the blast furnace are well documented. Although alkalis have unde-
sirable influences on coke strength and size in the furnace, different explanations have been invoked to account for losses in quality.
It is established that the alkali content and reactivity of coke
show remarkable increases during the passage of the coke through
the furnace.

The efficient operation of modern blast furnaces depends largely on the maintenance of high stock-column permeability. An important property of the feed coke, therefore, is that it remains physically stable in the chemical and thermal environment of the blast furnace to provide a permeable medium in the bosh region where the metallurgical coke is solid material [52].

Interaction of Potassium with Metallurgical Coke

Because the performance of coke in the blast furnace is influenced to a marked extent by "absorption" of alkalis, it is important to establish the mechanism by which alkalis modify the high-temperature properties of metallurgical coke.

The effect of potassium vapor, at temperatures between 900 and 1200°C, is to cause degradation of the coke particles through expansion (possibly intercalation), causing fissuring and eventual breakup of the coke particles.

Diffusion of the potassium is thought to occur preferentially within the more anisotropic components of the coke structure (Fig. 10).

Studies by Marsh et al. [53] of the reaction of coke surfaces with potassium carbonate or hydroxide in carbon dioxide at 1000°C indicate that the anisotropic carbon forms of larger sized optical texture (>10 µm) were oxidized and removed preferentially. This graphitizable carbon was more susceptible to attack by potassium salts than nongraphitizable (isotropic) carbon. The prismatic edges of the graphitizable carbon were more susceptible to alkali attack than the basal planes (Fig. 11).

Lu et al. [54] reported that the extents to which potassium "intercalated" into metallurgical cokes decreased with increased anisotropy. They found, in agreement with others [9], that the reaction of coke with potassium at 300-600°C, under nitrogen, led to the breakup of the coke to powder. The cracking of the coke was the result of concentration gradients of the potassium within the coke structure following "intercalation" in the less graphitic areas of graphitized cokes. They observed that porous inert particles did

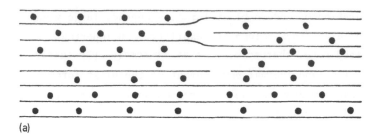

(a)

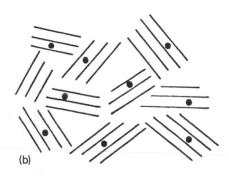

(b)

FIG. 10 Diagrams to illustrate the position of potassium ions (●)
in (a) the graphitizable (anisotropic) carbon and (b) the nongraphitizable (isotropic) carbon.

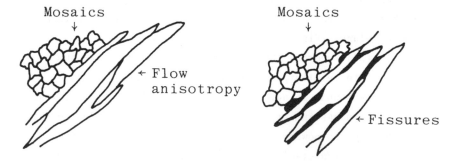

FIG. 11 Degradation of metallurgical coke in carbon dioxide by
potassium interaction within the flow-type anisotropy of optical
texture. (Left) Before intercalation. (Right) After intercalation.
(From Marsh et al. [53].)

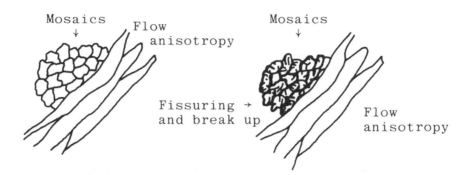

FIG. 12 Degradation of metallurgical coke in nitrogen by potassium interaction within the mosaics of optical texture. (Left) Before intercalation. (Right) After intercalation. (From Lu et al. [54].)

not change the cracking characteristics of the material. They observed more fissuring of the nongraphitizable components (mosaic optical textures) of cokes than of the graphitizable anisotropic components (flow anisotropy) (Fig. 12). With potassium vapor it is expected that, as the anisotropy of the coke is increased, so the amount of potassium *intercalated* will increase.

However, *degradation* is thought to be more pronounced in carbons with mosaic-type (1-10 μm size) optical textures. In this case, mixed coloring (yellow-blue-black) has been observed within the particles. This suggests that there is a nonhomogeneous distribution of potassium leading to large variations in expansion within the coke structure, causing high stresses and eventually crack formation, as shown in Fig. 13. This is supported by experiments of intercalation of potassium vapor into single-crystal graphite and metallurgical coke (HTT 1100°C) at 900°C. Preliminary results by Marsh et al. [55] indicated maximum degradation (breakup) occurred in the coke heat treated at 1100°C (mosaic optical texture) but the single-crystal graphite (HTT 3000°C) was relatively unaffected (Fig. 13) because single crystals have no spatial limitation to swelling. This may indicate that small amounts of potassium intercalated into carbons consisting of mosaic-optical textures have a more profound

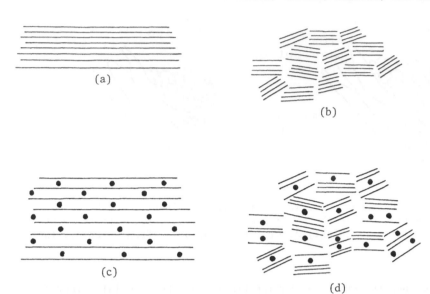

FIG. 13 Diagrams to illustrate the degradation of coke by nonsym-
metrical swelling within cokes: (a) anisotropic carbon (preintercalation); (b) mosaic optical textures (preintercalation), (c) potassium intercalation into graphitizable (anisotropic) carbon--expansion
can be accommodated (nondestructive); (d) mosaic optical textures
(after intercalation--expansions cannot be accommodated, and degradation of carbon pieces results).

effect (degradation) than larger amounts in a carbon consisting of
a flow-type optical texture. This may be due to ease of accommodation
of potassium atoms in the lattice of the more anisotropic (graphitizable) carbons (Fig. 13).

It is predicted that degradation of coke can be caused by the
expansion between the carbon layers, so creating fissures and breakup
of the original coke particles. It is anticipated that fissure development will be concentrated in the less anisotropic component of the
cokes as this is more susceptible to expansion caused by insertion of
the potassium, creating localized stresses and promoting fissure
development.

Chang and Tate [56] studied the degradation of coke by alkali
"adsorption" using two cokes made from ordinary and inferior coal
blends that had been used in an experimental blast furnace and had
been found to be problematic.

Results indicated that the rate of alkali adsorption at temperatures ranging from 800 to 1000°C under nitrogen depended on the initial porosity of the coke. Alkali was adsorbed by both cokes, and at 1 wt% uptake macrocracks were generated. Extents of macrocrack formation increased with increased alkali adsorption and were higher for the less porous cokes than for the more porous cokes at the same extents of alkali adsorption. Macrocrack formation, if caused by intercalation of potassium, would be expected to increase with decreased porosity of the carbon host. Potassium is very mobile in the intercalated state and gains access to the whole of the layered structures of the coke, the less porous coke being less able to accommodate internal stresses.

The presence of carbon dioxide in the atmosphere was found to diminish formation of macrocracks caused by alkali adsorption. Macrocrack formation was completely prevented at pressures of carbon dioxide of >5 kPa (0.05 atm). This is in line with the observations of Daumas and Herold [37] that intercalated potassium is oxidized by carbon dioxide above 120°C.

IV. ASSESSMENTS OF THE LITERATURE

A. Potassium

For single-crystal graphite materials, it is established that these materials readily form intercalation compounds with potassium and that these compounds totally decompose on heating to 600°C. Staging is well defined in such materials [2,3]. Industrial-grade graphites may behave similarly [4], and there is a report that a polycrystalline nuclear-grade graphite forms the KC_8 intercalate at 1000°C in the presence of potassium vapor [22].

For graphitizable carbons it is said that potassium forms intercalation compounds at reaction temperatures <500°C and that extents of incorporation of potassium into the carbons increase with increasing HTT of the carbons from 1000 to 3000°C [9,13]. A graphitizable petroleum coke, HTT 800°C, did *not* form intercalation compounds, but graphitizable carbons of HTT 1500°C gave x-ray diffraction patterns corresponding to KC_{12}-KC_{14} compounds.

For nongraphitizable carbons, it is said that potassium is adsorbed at pretreatment temperatures <500°C with compounds *not* formed [6]. For vitreous and saccharose carbons, HTT 1000-2500°C, reacting with potassium vapor at temperatures reaching 800°C, relatively large quantities of potassium in the intercalated state are stable to the higher reaction temperatures [9]. The volumes of potassium taken up by the carbons exceeded the pore volume of the carbon, indicating that swelling or imbibition had occurred [9]. Intercalation has been reported for reaction of potassium from potassium carbonate with an active carbon at 827°C [24].

Thus, an overall trend for reactions of carbons with potassium is that single-crystal graphite forms structured intercalation compounds having well-defined diffraction patterns that can be interpreted in terms of staging. For the graphitizable but nongraphitic carbons of low internal surface area (probably <20 $m^2 g^{-1}$), it has been reported that the less ordered carbons of HTT <800°C do not take up potassium. With such carbons of increasing HTT and associated enhanced lamellar and stacking perfection, extents of potassium uptake decrease. Also, the thermal stability of the carbon-potassium systems increase beyond those of the graphite-potassium systems. For the more disordered nongraphitizable carbons, usually of relatively high internal surface areas (100-1000 $m^2 g^{-1}$), the uptake of potassium is considerable, is apparently independent of HTT, and has the highest thermal stability.

B. Sodium

Intercalation of sodium in well-crystallized natural and synthetic graphites has been described [7,9,57]; only higher stages were observed: stage 8 [7,9] or stages 4, 6, and 7 [57], respectively. Graphitizable carbon at 600°C was reported to react with sodium, the extents of uptake of sodium decreasing with increasing HTT of the carbon, 1250-2000°C [9]. It is suggested that this uptake of sodium is an adsorption process [7], although this has been disputed [47].

Carbon blacks behave similarly [47]. Sodium appears to be taken up by the microporous nongraphitizable carbons, and there is some evidence for intercalation [9,47].

C. Assessments

Thus, potassium appears to be able to enter into all carbon forms except possibly graphitizable carbons of low HTT. Sodium will intercalate graphite and appears to be able to enter into both graphitizable and nongraphitizable carbons. Detailed information is not available. For the single-crystal graphites, the graphite-potassium system would exhibit a maximum in structural ordering with the well-defined staging of intercalation compounds in which the driving force for intercalation is the free energy change associated with electron donation from potassium to the graphite. With the less ordered graphitizable and nongraphitizable carbons, the same type of reaction can occur, but because of the presence of various kinds of structural defects, the identification of intercalation compounds with well-defined regular staging cannot be expected. The technique of x-ray diffraction is therefore not appropriate for a full characterization of these systems. However, swelling apparently occurs, and incorporation of potassium or sodium metal, as *atoms*, as distinct from *ions*, also appears possible. The extent of charge transfer in such compounds as KC_8 is still a matter of discussion. In higher stage compounds, potassium seems to be more ionic than in stage 1 compounds. The facility of access within microporosity of the nongraphitizing carbons, as well as adsorption of potassium and sodium, is an additional factor.

The potassium compound is stable only in vacuum or an inert atmosphere and very readily decomposes in the presence of CO_2, O_2, and H_2O. Because of these important properties, it is difficult to assume that the mechanism of intercalation solely causes degradation or comminution of metallurgical coke in the blast furnace. Similarly, studies that react carbon with potassium carbonate in an atmosphere of carbon dioxide are difficult to interpret because of the additional

complexity of rapid catalytically induced gasification of carbon
forms. Under these conditions, it is not certain whether oxides of
potassium can enter into carbon structures or whether there is
mobility of potassium (oxides) over and through carbon forms to produce gasification channeling or fissures. Certainly, in these systems, it appears that the graphitizable (anisotropic) components of
a metallurgical coke are more reactive than the nongraphitizable
(isotropic) components.

It has been predicted [18] that, in the presence of carbon
dioxide, potassium intercalation compounds are unstable below 1127°C.

The uptake of potassium or sodium by both types of carbon causes
swelling. If the carbon piece is relatively large or if a carbon
composite material is used that contains several carbon forms, as in
metallurgical coke, then differential swelling effects can place
intolerable strains within the carbon piece and disintegration or
fracture can occur [57].

V. POTASSIUM INTERCALATION: UNDERSTANDING THE MECHANISM

A. Preparation of Potassium Intercalation Compounds

Several methods for the preparation of graphite intercalation compounds have been developed [58], including two-zone vapor transport,
liquid intercalation, and electrochemical and cointercalation techniques. Controlling parameters are temperature, vapor pressure, the
chemical and physical properties of the intercalate, and the characteristics of the carbon or graphite host.

Intercalation compounds can be prepared from solid, liquid, or
gaseous reagents [52], the most common method being by vapor transport using the two-zone method.

B. Two-Zone Vapor Transport

In this method, the intercalate (potassium) is heated to temperature
T_i in one ampoule and passes, as vapor, to a second ampoule containing the carbon at a higher temperature T_c (Fig. 2). The stage of the
intercalation compound and/or the weight uptake of intercalate is

controlled by the chemical potential of the potassium and depends upon p/p^0 (p^0 = vapor pressure at T_c and p = vapor pressure at T_i). The smaller the difference between T_c and T_i, the lower is the stage of the intercalation compound formed.

Generally, for potassium, the temperature of the metal is kept constant while the temperature of the carbon is varied. In this way the stages can be controlled. The first-stage compound of potassium-graphite is best developed when the temperature difference between the graphite and potassium is <60°C. The reaction temperature T_c and associated stages of intercalation for potassium compounds [10,11] are shown in Table 6. Table 6 and Fig. 3, showing the isobars for the formation of the potassium-graphite compounds [10], indicate that a given stage can be prepared over a range of reaction temperatures T_c.

Higher staged compounds can also be prepared using the two-zone vapor transport technique. However, the staging depends on several parameters, such as geometric factors, sample size (thickness and cross-sectional area), ampoule size and shape, and the accuracy of temperature control. In the preparation of dilute alkali metal compounds, the shaping of the ampoule to minimize the free volume for intercalate vapor around the graphite and "necking down" the constriction between the graphite and the intercalate zones leads to an

TABLE 6 Potassium-Graphite Intercalation Compounds (Potassium Temperature T_i = 250°C)

Stage	Formula	Reaction temperature T_c of graphite (°C)	Color	Reference
1	KC_8	225-320	Yellow-gold-red	3
2	KC_{24}	350-400	Blue	3
3	KC_{36}	450-480	Blue-black	3
4	KC_{48}	—	Black	4
5	KC_{60}	—	Black	4
6	—	500	—	3

increase in stage index of the sample [38]. The intercalation of small graphite samples promotes the formation of a single stage throughout the sample volume.

C. Structure of Alkali Metal Intercalation Compounds

The first determination of the structure of potassium-graphite was carried out by Schleede and Wellmann [2]. The introduction of the alkali metal atoms increased the interplanar distances from 335.4 pm in a highly orientated pyrolytic graphite to 541 pm for potassium-graphite.

In the first-stage compound, alkali metal atoms form a triangular net, with a length edge of 491 pm, that is, exactly twice the a-dimension of the hexagonal carbon net. When the layers are fully occupied, the first-stage compound has the composition KC_8.

A later refined model for the above compound [3] is shown in Fig. 14. It differs from the model of Schleede and Wellmann [2] in orientation of the carbon and alkali metal planes relative to one another. All the carbon planes are in identical positions (i.e., they have a sequence AAA···), the alkali metal atoms occupying positions above and below the middle of a hexagonal ring of carbon atoms. In this arrangement every metal atom has 12 carbon atoms at the same distance from it (307 pm for KC_8).

As the metal layers in the second-stage compound are separated by two carbon layers, the C/K ratio should be twice that of first-stage compounds, that is, KC_{16}. This is not so. The composition was KC_{24}, the metal atoms being less densely packed. It was assumed at first that every third K atom is removed from the triangular network, resulting in a hexagonal network. It is now believed that the potassium layers are disordered and can be considered a lattice gas [59]. At very low temperatures several ordered structures have been observed, commensurate and noncommensurate with the honeycomb pattern of the carbon sheets.

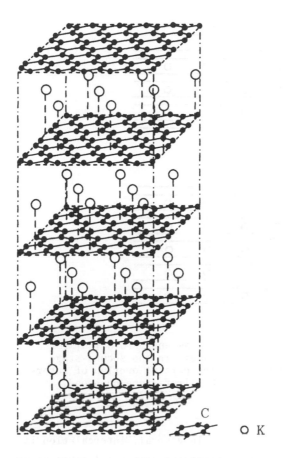

FIG. 14 Structure of potassium-graphite KC_8. (From Ref. 3.)

D. Staging Phenomena and Phase Transitions

The ability to form stages is one of the most characteristic properties of graphite intercalation compounds. Staging is the formation of an ordered array of alternate graphite and intercalated layers, in which the stage number n designates the number of graphite layers between any two intercalated layers.

The phenomenon of staging has been known for a long time [2,3]. However, only recently has it attracted considerable experimental as

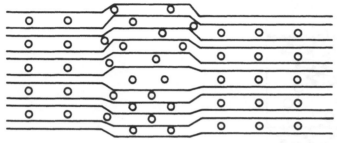

(a)

(b)

FIG. 15   (a) Domain model proposed by Daumas and Herold [60] for
stacking of graphite and alkali metal layers in the second-stage
compound.   (b) Possible structure showing partial overlap of intercalated domains on transformation to the first-stage compound.
(From Ref. 63.)

well as theoretical research activity. In general, intercalation is
readily achieved, and the stage transformation n to n ± 1 takes place
with ease. The range of order is usually quite long, extending well
over 100 nm in directions both perpendicular and parallel to the
layers.

Daumas and Herold [60] proposed a staging model that assumes
domains of intercalated material between graphite planes. These
domains are stacked in columns to form ordered stages (see Fig. 15a).
Stage transformations merely require a spreading out and rearrange-
ment of the islands without the necessity for occupied intercalated
layers to be completely evacuated while others are filled [59].
This model has been used to explain why x-ray diffraction patterns
taken on a sample during intercalation show the sample to pass
through a sequence of stages.

Dowell and Badorrek [61] have argued that the absence of a change in diffusion coefficient during a stage transformation provides strong support for the Daumas-Herold domain model because this model does not require extensive intercalate migration to produce a stage transformation.

McRae et al. [62] noted that, for stage 15 graphite-potassium, the electrical resistivity is three to four times lower than that of graphite, indicating that intercalation corresponds to an ordered state, not merely to a random addition of intercalate.

Misenheimer and Zabel [63] report an x-ray investigation of the immediate transition region of the stage transformation n to n ± 1. According to the Daumas-Herold model, during the transformation the intercalated islands may partially overlap, causing stacking disorder and/or large coherency stresses at the edges of the intercalated islands (Fig. 15b). Disorder and stresses, in turn, should lead to diffuse scattering and broadening of Bragg reflections.

X-ray measurements were carried out on potassium compounds from highly oriented pyrolytic graphites prepared in situ in a two-zone furnace mounted on an x-ray stage. The authors concluded that the stage transformations were predominantly controlled by the nucleation of domains rather than by the diffusion-dependent growth kinetics of the alkali islands [63].

The intercalate islands postulated by the Daumas-Herold model have been observed by lattice imaging TEM (for $FeCl_3$-graphite) [64].

Hooley et al. [33] showed that no intercalation occurs on exposure of pyrographite to bromine vapor when the basal faces of the graphite are protected by a cover. These and other kinetic studies, summarized in a review [59], indicated that intercalation starts from the interlayer spaces nearest the basal faces and proceeds sequentially into interspaces increasingly distant from the basal faces. The reason seems to be that the outermost carbon sheets are the most flexible, able to yield to the mechanical stresses of intercalation [65].

Molecular alignment studies in liquid crystals by de Gennes [66] show that the elastic force field decreases the strain energy (the separation energy to admit the intercalated substances into the graphite) by clustering and aligning neighboring molecules. Application of this concept to graphite intercalation compounds implies that the strain energy is minimized by clustering of the intercalate to form platelets between a single set of graphite planes in which intraplanar forces are extremely strong. Thus, if a random distribution of intercalated atoms were introduced into graphite initially, the strain energy of the system would be lowered by the clustering of all intercalated atoms between a single set of graphite planes. However, each intercalate layer can accommodate a limited number of atoms defined by the close-packed condition. Therefore, the minimum strain energy is achieved by the closest packing within the intercalate layer and by the largest separation between intercalate layers consistent with a given intercalate concentration.

Safran and Hamaan [67,68] considered the concept of strain energy an important factor for staging in graphite intercalation compounds according to the Daumas-Herold domain model. These authors calculated the strain-induced forces between two intercalated atoms in a layered graphite host material. They showed that intercalated atoms between the same two graphite layers are attached to one another to form two-dimensional islands, and intercalated atoms on different layers interact repulsively. By forming intercalate islands, the compound lowers its strain energy by ~0.1 eV per intercalated atom, and the free energy is minimized by the formation of a pure stage configuration from a mixed or randomly staged crystal. The arrangement of the intercalate islands in a mixed-stage compound is illustrated in Fig. 16b, the strain being localized in the boundary region between adjacent islands.

Nishitani et al. [69] determined a precise phase diagram for stages of the potassium-graphite intercalation compounds. They used in situ x-ray diffraction to elucidate the origin of staging using highly oriented pyrolytic graphite with potassium metal. For potassium-graphite intercalation compounds, they observed no evidence of a disordered mixture of two stages, in agreement with Ref. 63.

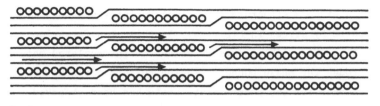

(a)

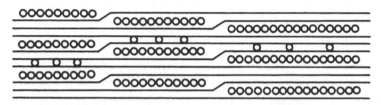

(b)

3rd stage   2nd stage

(c)

FIG. 16 (a) Daumas-Herold model for the third-stage compound.
(b) Possible intermediate structure in the transformation from
third- to second-stage compound. (c) Separation of second- and
third-stage compound during the transition from third- to second-
stage compound. (From Ref. 69.)

The mixed staging observed in graphitizable carbons is compatible
with these studies if the large variation in crystallite structure
or perfection is considered.

They considered a stage transformation as intercalate diffusion
on the basis of the Daumas-Herold model. A stage n structure has n
different domains (Fig. 16a). For instance, a stage 3 structure can

transform into a stage 2 structure (Fig. 16c) by intercalating diffusion, as indicated by arrows in Fig. 16a. In this transformation, an
intermediate state (Fig. 16b) occurs in which the intercalated material passes through the phase boundary region to form the domains.
However, no evidence for the intermediate state has been observed
during the stage transformation.

The authors concluded that, in the stage transformation, either
the stage boundary region should be small compared with the remaining
well-staged regions or the diffusion of intercalated alkali metal
should be rapid in this region compared with the dwelling period at
the domains.

It is not clear at the present time that the Daumas-Herold model
can, by itself, fully account for all aspects of intercalation and
staging, although it goes a long way to doing so.

VI. CONCLUSIONS

The interactions of graphitizable, nongraphitizable, and graphitic
carbons with potassium and sodium have been reviewed.

The overall conclusions of this review are summarized in Tables
7-12 and indicate the strength of available evidence for intercalation
formation during interactions of potassium and sodium with the many
forms of carbon. Tables 7-12 give the conditions of reactions, the
experimental techniques used, and comment. In summary, we can conclude the following.

1. Potassium forms intercalation compounds with graphitic
    carbons at temperatures of 300-500°C under vacuum or an
    inert gas. The interlayer spacing increases from 335.4 to
    541 pm in the intercalate. Intercalation produces charac-
    teristic changes in light adsorption. These are seen as
    color changes in the lower stages: yellow-gold, first
    stage: KC_8; blue, second stage: KC_{24}; blue-black, third
    stage: KC_{36}; black, fourth stage: KC_{48}. These stages are
    dependent on the temperature and vapor pressure of the
    potassium (Table 7).

TABLE 7 Summary of Potassium Reactions with Graphites

Reaction temperature (°C)	Graphitic carbon with potassium[a]
300-500	Potassium metal + 1, 2 (under N_2 vacuum) [1,4]
	X-ray diffraction and isobar curves
	KC_8, yellow-gold, first stage
	KC_{24}, blue, second stage
	KC_{36}, blue-black, third stage
	Interlayer spacing increased from 335.4 to 541 pm
	Intercalation[b]
800	Potassium + 2 (under N_2) [24]
	X-ray diffraction, change in (00ℓ) intercalation structures formed
	No intercalation
1000	Potassium + 2 (under Ar) [22]
	X-ray diffraction, change in (00ℓ)
	1000°C graphite, KC_8 formed
	1200°C graphite, KC_{60} formed
	Intercalation[c]

[a]Carbon type. (1) natural flake graphite; (2) artificial graphite.
[b]Conclusive evidence.
[c]Questionable evidence.

2. Experimental evidence using K_2CO_3 or K metal under nitrogen at elevated temperatures (600-1000°C) is both limited and conflicting. Potassium metal appears to form graphite compounds, but K_2CO_3 does not. Thermodynamic considerations indicate that, in the presence of carbon dioxide, intercalation compounds are not formed at temperatures <1127°C. However, in the absence of carbon dioxide, intercalation compounds should form at temperatures of about 1000°C (Table 7).

3. Although sodium intercalation compounds with graphite have been reported, it would appear that low-staged (first

TABLE 8   Summary of Sodium Reactions with Graphitic Carbons

Reaction temperature (°C)	Graphitic carbons with sodium[a]
400	Sodium + 1c + 2c under vacuum [45]
	X-ray diffraction and magnetic susceptibility change in diffraction pattern for 1 and 2, disappearance of 0002 graphite line indicates nearly all graphite has reacted in some way; new lines appear near site of 0002 graphite line owing to new compound formation; lines $10\bar{1}0$, $11\bar{2}0$, $20\bar{2}0$, and $12\bar{3}0$, no change indicates structure of graphite sheets unaffected, that is, lamellar; magnetic susceptibility changes from strong diamagnetism to slight paramagnetism
	Intercalation[b] (impurities can affect this reaction)
1000	Sodium + pure graphite [9]
	X-ray thermogravimetric analysis; no change recorded
	No intercalation[b]

[a]Carbon type: (1) natural graphite, (2) artificial graphite; grain size (c), <1 nm.
[b]Conclusive evidence.

through fourth) intercalation compounds are not formed with pure graphite. Impurities within the graphite-sodium system, such as boron, or defects may enhance the ability of intercalate formation (Table 8).

4. Potassium forms intercalation compounds with graphitizable carbons heat treated above HTT 1000°C under vacuum or nitrogen at temperatures below 500°C. The extent of intercalation increases with the HTT of the carbon. No compounds have been observed with graphitizable carbons of HTT <800°C. X-ray diffraction indicates that mixed staging is characteristic of graphitizable carbons of HTT <1500°C (Table 9).

5. Intercalatelike structures can form in graphitizable carbons at reaction temperatures 800-1000°C under vacuum or nitrogen. A carbon dioxide atmosphere prohibits the formation of intercalates up to temperatures of at least 1127°C (Table 9).

TABLE 9 Summary of Potassium Reactions with Graphitizable Carbons

Reaction temperature (°C)	Graphitizable carbons with potassium
300	Potassium + 3c[a] (HTT 1250, 1600, 2000, and 2400°C under vacuum) [9]
	Thermogravimetric analysis and x-ray study; new band formed between (002) and (001): the (10)$_K$ band caused by two-dimensional potassium network (HTT 2000 + 2400°C); diffuse band between (002) and (001) in 1250 and 1600°C; amount of intercalation increases with increasing HTT
	Intercalation[b]
	Potassium + 3c[a] (HTT 800, 1000, 1500, 2300°C) [13]
	X-ray diffraction and gravimetric analysis; HTT 1000-1500°C (suggests wide distribution of interlayer spacings); HTT 2300°C x-ray profile (sharp; single staging); HTT 800°C, no intercalates formed
	Intercalation compounds require HTT > 800°C[b]
1000	Potassium + coke (under CO_2) [18]
	Thermodynamic considerations
	No compounds form in CO_2 atmosphere up to temperature > 1120°C
1200	Potassium + 3 + 4[a] under Ar/CO [22]
	X-ray analysis: 3(CO)-KHCO$_3$, 3(Ar)K$_2$CO$_3$, 4(C)-KC$_8$ to KC$_{24}$; 4(Ar)KC$_{24}$
	Intercalates possible[c]

[a]Carbon type: (3) petroleum cokes, (4) metallurgical cokes. Grain size c < 1 mm.
[b]Conclusive evidence.
[c]Questionable evidence.

6. Sodium may form intercalation compounds with graphitizable carbons (HTT not higher than 1600°C) under vacuum or inert gas at 600°C. Intercalation may be accompanied by adsorption in micropores. Sodium uptake decreases with increasing HTT of the carbon and is a maximum for a carbon of HTT 1000°C (Table 10).

TABLE 10   Summary of Sodium Reactions with Graphitizable Carbons

Reaction temperature (°C)	Graphitizable carbon with sodium
400	Sodium + 3[a] (HTT 1250, 1600, 2000, and 2400°C) [9]
	Thermogravimetric and x-ray analysis; with increased HTT the uptake of sodium decreased: graphitic carbons do not fix sodium; x-ray analysis indicated a noticeable shift in (002) for 1250 and 1600°C HTT cokes; an increase in parameter c[b] and parameter a[b] is unchanged: intercalation compounds with limited adsorption[c]
	Intercalation in 1250 and 1600°C HTT; adsorption in 2000 and 2400°C HTT
1000	Sodium + 3[a] (HTT 1100°C under vacuum)
	Gravimetric uptake of sodium; isobar curves indicative of adsorption[c]
	Adsorption

[a]Carbon type: (3) petroleum coke.
[b]Grain size (mm): (a) > 5, (b) > 1, (c) < 1.
[c]Conclusive evidence.

7.   Potassium intercalation with nongraphitizable carbons have been reported. However, the evidence is not conclusive. Adsorption effects within micropores of these carbons, causing swelling, cannot be ruled out and could produce the effects observed although it may be only a matter of interpretation at this level of order (Table 11).

8.   Sodium appears to form intercalation compounds at 600°C with nongraphitizable carbons of HTT 1000-1500°C but not of 2000-2500°C under vacuum or inert gas. However, adsorption within the micropores, causing swelling, could produce the effects observed (Table 12).

9.   Very little (if any) of the literature is directly comparable. The heterogeneity of coals and pitches and their treatments provide a wide variety of graphitizable carbons with differing properties. The variety of graphites in

TABLE 11 Summary of Potassium Reactions with Nongraphitizable Carbons

Reaction temperature (°C)	Nongraphitizable carbons with potassium
300	Potassium + 5c[a] (HTT 1000 and 2100°C under vacuum) [7]
	Gravimetric analysis—profiles characteristic of adsorption
	No intercalation[b]
	Potassium + 5 (HTT 1000, 1500, 2000, and 2500°C), 6 (HTT 1100°C), 7(HTT 1000 and 2500°C) [9]
	Thermogravimetric and x-ray analysis: potassium uptake accessible porosity; intercalate;[b] x-ray profiles diffuse, possible $(10)_K$ band
	Intercalation[b]
800	Potassium carbonate + carbon + activated carbon (HTT 1000°C under N_2) [24]
	X-ray analysis: profile similar to KC_{36}-graphite[b]
	Intercalation[b]
1000	No information:
	Potassium metal, K_2CO_3 + C 8a[a] under nitrogen [55]
	Physical examination, x-ray
	No intercalation[b]

[a]Carbon type: (5) sugar carbons saccharose, (6) polyvinylidene chloride coke, (7) glassy carbons, (8) carbon black, (9) 901 coal char. Grain size (mm): (a) >5; (b) >1; (c) <1.
[b]Questionable evidence.

terms of their mode of formation (natural or synthetic), heat treatment temperature, and level of impurities make direct comparisons between intercalation studies problematic.

10. There is little information outlining the variations with temperature of stabilities of compounds in the presence of potassium vapor as distinct from a vacuum or inert gas.

11. The stability of potassium incorporation (intercalation?) at elevated temperatures (~800-1200°C) in graphitizable but not graphitic carbons appears to be higher than the stabilities of potassium incorporated in graphitic carbons.

TABLE 12   Summary of Sodium Reaction with Nongraphitizable Carbon

Reaction temperature (°C)	Nongraphitizable carbon with sodium
400	Sodium + 5[a] (HTT 1000 and 2100°C under vacuum) [6]
	Gravimetric analysis: profiles characteristic of adsorption
	No intercalation: adsorption into microporosity only
600	Sodium + 5 (HTT 1000, 1500, 2000, and 2500°C), 6 (HTT 1100°C), 7 (HTT 1000 and 2000°C) under vacuum [9]
	Thermogravimetric and x-ray analysis: 2000 and 2500°C; uptake of sodium indicative of adsorption; no evidence of swelling; x-ray (2000 + 2500°C HTT) analysis indicates slight broadening of (002) band: sodium does not intercalate these carbons
	Adsorption only in 2000 and 2500°C HTT; x-ray (1000, 1100, and 1500°C HTT) 5, 6, 7; shift of the (00ℓ) indicative of intercalation[b]
	Nongraphitizable carbons HTT 1000-1500°C may form intercalation compounds

[a]Carbon type: (5) saccharose carbon, (6) polyvinylidene chloride, (7) glassy carbon.
[b]Questionable evidence.

12. Potassium is probably adsorbed by isotropic nongraphitizable carbons. Adsorption or imbibition may cause expansion and degradation of the carbon. This is an area of study that has not received much attention. Potassium may be a "unique" absorbate and may also be incorporated into isotropic structures.

13. Potassium compounds are unstable in the presence of oxidizing gases (H_2O, CO_2, and O_2) at temperatures up to ~1127°C, for *graphitizable* and *graphitic* carbons. Hence, the study of intercalation must be made in oxygen-free, dry environments:

$$\text{Potassium - } \begin{matrix} \text{intercalation} \\ \text{compound} \end{matrix} + CO_2 \rightarrow KO_x + \text{carbon} + CO$$

14. The equilibrium position and rates of this reaction must be considered in order to assess the ability of potassium to be incorporated within the metallurgical cokes of the blast furnace.

15. The studies of the Northern Carbon Research Laboratories indicate that the smallest sizes of optical texture are *least* resistant to disintegration by potassium vapor, (e.g., fine-grained mosaics). The largest sizes of optical texture appear to be most resistant to reaction by alkali but are unsuitable in metallurgical cokes because of reductions in strength.

Metallurgical coke within the blast furnace is able to withstand thermal and mechanical shock by reason of its composite structure of anisotropic components of size, ranging from mosaics (1-10 μm diameter) to flow-type anisotropy (>10 μm). The mosaics contribute to strength and the flow-type anisotropy, with its elongated fissuring, can accommodate thermal expansions. Resistance to alkali attack by metallurgical cokes appears to be greatest with flow-type anisotropy, but a coke entirely of such structure would have unsuitable strength characteristics.

Because of these conflicting factors in the overall suitability of coke for the blast furnace, a coke structure with a dominance of coarse-grained mosaics (5-10 μm diameter) appears to be a necessary compromise.

ACKNOWLEDGMENTS

The authors are sincerely grateful for the financial assistance provided by the European Coal and Steel Commission of the European Economic Community under grant 7220-EB/821. The assistance with the preparation of the manuscript by Mrs. Marion Poad, Miss Bridget A. Clow, Mrs. Patricia M. Wooster, and Mrs. Elaine Watson is acknowledged.

REFERENCES

1. K. Fredenhagen and G. Cadenbach, Die Bindung von Kalium durch Kohlenstoff, *Z. Anorg. Allg. Chem. 158-9,* 249 (1927).

2. A. Schleede and M. Wellmann, Über die Struktur der Einwirkungsprodukte von Alkalimetallen auf Graphit, *Z. Phys. Chem. B18,* 1 (1932).

3. W. Rudorff and R. Schulze, Über Alkali-Graphit Verbindungen, *Z. Anorg. Allg. Chem. 277,* 156 (1954).

4. A. Herold, Recherches sur les composés d'insertion du graphite, *Bull. Soc. Chim. Fr.,* 999 (1955).

5. R. E. Franklin, Crystallite growth in graphitizing and non-graphitizing carbons, *Proc. R. Soc.* (London) *Ser. A. 209,* 196 (1951).

6. U. Hofmann, A. Ragoss, G. Rüdorff, R. Holst, W. Ruston, A. Russ, and G. Ruess, Die Struktur and die Graphitierung von Kohlenstoff, *Z. Anorg. Allg. Chem., 255,* 195 (1947).

7. N. Platzer-Rideau, Quelques propriétés des carbones graphitables et non-graphitables, *Ann. Chim.* (Paris) *51,* 805.060 (1960).

8. B. Caston and A. Herold, Sur la réaction du potassium avec le carbone vitreux, *C. R. Acad. Sci., Ser. C 269,* 792 (1969); Recherches sur les systemes carbones—potassium. II. Étude des produits de réaction, *Bull. Soc. Chim. Fr.,* 521 (1972).

9. D. Berger, B. Carton, A. Metrot, and A. Herold, Interactions of potassium and sodium with carbons, in *Chemistry and Physics of Carbon,* Vol. 12, P. L. Walker, Jr., and P. A. Thrower, eds., Marcel Dekker, New York, 1975, p. 1.

10. A. R. Ubbelohde and F. A. Lewis, *Graphite and Its Crystal Compounds,* Clarendon Press, Oxford, 1960.

11. D. E. Nixon and G. S. Parry, The formation and structure of potassium-graphites, *Br. J. Appl. Phys. D1,* 291 (1968).

12. C. Herinckx, R. Perret, and W. Ruland, Interstitial compounds of potassium with carbon fibers, *Nature 220,* 63 (1968); Interstitial compounds of potassium with carbon fibers, *Carbon 10,* 711 (1972).

13. T. Terai and Y. Takahashi, Sorption behaviour of hydrogen gas and tritiated hydrogen gas on potassium-carbon intercalation compounds, *Carbon 22,* 91 (1984).

14. P. Kwizera, M. S. Dresselhaus, and G. Dresselhaus, Raman spectra and staging of intercalated graphite fibers, *Carbon 21,* 121 (1983).

15. F. Huhn, J. Klein, and H. Juntgen, Investigations on the alkali-catalysed steam gasification of coal: Kinetics and interactions of alkali catalyst with carbon, *Fuel 62*, 121 (1983).

16. L. L. Ban, D. Crawford, and H. Marsh, Lattice-resolution electron microscopy in structural studies of non-graphitizable carbons from polyvinylidene chloride (PVDC), *J. Appl. Crystallogr. 8*, 415 (1975).

17. G. R. Hennig, Intercalation compounds of graphite, *Prog. Inorg. Chem. 1*, 125 (1959).

18. Y. K. Rao, A. Adjorlolo, and J. H. Haberman, On the mechanism of catalysis of the Boudouard reaction by alkali-metal compounds, *Carbon 20*, 207 (1982).

19. D. A. Fox and A. H. White, Effect of sodium carbonate upon gasification of carbon and production of producer gas, *Ind. Eng. Chem. 23*, 259-266 (1931).

20. D. W. McKee, Mechanisms of the alkali metal catalysed gasification of carbon, *Fuel 62*, 170 (1983).

21. B. J. Wood, R. D. Brittain, and K. H. Lau, A study of the role of alkali metal salts as char gasification catalysts by Knudsen cell mass spectroscopy, *Carbon 23*, 73 (1985).

22. R. J. Hawkins, L. Monte, and J. J. Waters, Potassium attack of blast furnace refractory carbons, *Ironmaking Steelmaking 1*, 151 (1974).

23. S. Yokoyama, K. Tanaka, I. Toyoshima, K. Miyahara, K. Yoshida, and J. Tashiro, X-ray photoelectron spectroscopic study of the surface of carbon doped with potassium carbonate, *Chem. Lett.*, 599 (1980).

24. F. Kapteijn, J. Jurriaans, and J. A. Moulijn, Formation of intercalate-like structures by heat treatment of K_2CO_3-carbon in an inert atmosphere, *Fuel 62*, 249 (1983).

25. H. E. Podall, W. E. Foster, and A. P. Giraitis, Catalytic graphite inclusion compounds. I. Potassium graphite as a polymerization catalyst, *J. Org. Chem. 23*, 82 (1958).

26. E. Ferguson, R. Schlogl, and W. Jones, Gasification of potassium intercalated and impregnated natural graphites, *Fuel 63*, 1048 (1984).

27. T. Wigmans, J. C. Goebel, and J. A. Moulijn, The influence of pretreatment conditions on the activity and stability of sodium and potassium catalysts in carbon-steam reactions, *Carbon 21*, 295 (1983).

28. W. Y. Wen, Mechanisms of alkali metal catalysis in the gasification of coal, char or graphites, *Catal. Rev. Sci. Eng. 22*, 1 (1980).

29. S. Aronson, F. J. Salzano, and D. Bellafiore, Thermodynamic properties of potassium-graphite lamellar compounds from solid-state EMF measurements, *J. Chem. Phys. 49*, 434 (1968).

30. J. A. Moulijn, M. B. Cerfontain, and F. Kapteijn, Mechanism of the potassium catalysed gasification of carbon in CO_2, *Fuel 63*, 1043 (1984).

31. B. J. Wood and K. M. Sancier, The mechanism of the catalytic gasification of coal char: A critical review, *Catal. Rev. Sci. Eng. 26*, 233 (1984).

32. B. J. Wood, R. H. Fleming, and H. Wise, Reactive intermediate in the alkali-carbonate-catalysed gasification of coal char, *Fuel 63*, 1600 (1984).

33. J. G. Hooley, W. P. Garby, and J. Valentin, The effect of sample shape on the bromination of graphite, *Carbon 3*, 7 (1965).

34. A. R. Ubbelohde, Electrical properties and phase transformations of graphite nitrates, *Proc. R. Soc.* (London) *Ser. A 304*, 25 (1968).

35. M. B. Dowell, Influence of graphite structure on rates of intercalation, *Mater. Sci. Eng. 31*, 129 (1977).

36. S. Kobayashi and Y. Omari, Rate of potassium uptake of coke from molecular nitrogen-potassium gas mixtures, *Tetsu to Hagane* (Japan) *66*, 1791 (1980).

37. N. Daumas and A. Herold, Propriétés chimiques des composés d'insertion due graphite: Action de l'oxygene et des composés oxygenes volatils sur les composés graphite-potassium, *Bull. Soc. Chim. Fr.*, 1598 (1971).

38. C. Underhill, T. Krapchev, and M. S. Dresselhaus, Synthesis and characterization of high-stage alkali metal donor compounds, *Synth. Met. 2*, 47 (1980).

39. N. Akuzawa, T. Fujisawa, and T. Amemiya, Stability of potassium-graphite intercalation compounds in an oxygen atmosphere, *Synth. Met. 7*, 57 (1983).

40. R. Schlogl and H. P. Boehm, The reaction of potassium-graphite intercalation compounds with water, *Carbon 22*, 351 (1984).

41. D. E. Bergbreiter and J. M. Killough, Lewis base properties of potassium-graphite, *J. Chem. Soc. Chem. Commun.*, 913 (1976).

42. L. B. Ebert, L. Matty, D. R. Mills, and J. C. Scanlon, The interrelationship of graphite intercalation compounds, ions of aromatic hydrocarbons and coal conversion, *Mater. Res. Bull. 15,* 251 (1980).

43. A. Metrot and A. Herold, Insertion et adsorption de sodium par les carbones, *J. Chim. Phys. Physicochim. Biol.,* 71 (1969).

44. M. C. Robert, M. Oberlin, and J. Mering, Lamellar reactions in graphitizable carbons, in *Chemistry and Physics of Carbon,* Vol. 10, P. L. Walker and P. A. Thrower, eds., Marcel Dekker, New York, 1973, p. 141.

45. R. C. Asher, Lamellar compound of sodium and graphite, *J. Inorg. Nucl. Chem. 10,* 238 (1959).

46. R. C. Asher and S. A. Wilson, Lamellar compound of sodium with graphite, *Nature 181,* 409 (1958).

47. I. Pflugmacher and H. P. Boehm, Intercalation of sodium by carbon blacks, *Proceedings of the Third Conference on Industrial Carbons and Graphites 1970,* Society of Chemical Industry, London, 1971, p. 62.

48. P. Albert and J. Parisot, Influence de certains catalyseurs de graphitation sur les propriétés du graphite, en particulier, sur les conductibilitiés électrique et thermique, *Proceedings of the Third Conference on Carbon, 1957,* Pergamon Press, New York, 1959, p. 467.

49. J. Davies, J. T. Moon, and F. B. Traice, Alkalis in the blast furnace, *Ironmaking Steelmaking 5,* 151 (1978).

50. J. Willems, G. Heynert, G. Quade, and W. Zischkale, Untersuch-ungen über Ansätze in Hochofen und ihre Beseitigung durch Spren-gen, *Stahl Eisen 84,* 57 (1964).

51. H. Itaya, F. Aratani, T. Funakoshi, A. Kani, and S. Kiyohra, Estimation of temperature profile, radial distribution of descending rate of burden material and ore/coke in the blast furnace, *Trans. Iron Steel Inst. Japan 20,* 199 (1980).

52. J. Goleczka, J. Tucker, and G. Everett, private communication, 1983.

53. H. Marsh, A. Wilkinson, and J. Machnikowski, Study of the reac-tion of cokes with KOH under nitrogen, *Fuel 61,* 834 (1982).

54. W. K. Lu, G. Samaan, and M. Uribe, Alkalis, textures of carbon and the degradation of coke in the blast furnaces, *Iron Steel-Maker 8,* 24 (1981).

55. H. Marsh, N. Murdie, and M.-F. Lin, unpublished results, 1984.

56. T. Chang and M. Tate, Degradation of coke due to gasification, *Trans. Iron Steel Inst. Japan, 19,* 479 (1979).

57. A. Metrot, D. Guerard, D. Billaud, and A. Herold, New results about the sodium-graphite system, *Synth. Met. 1,* 363 (1980).

58. A. Herold, Reflexions sur la synthese des composes lamellaires, *Mater. Sci. Eng. 31,* 1 (1977).

59. M. S. Dresselhaus and G. Dresselhaus, Intercalation compounds of graphite, *Adv. Phys. 30,* 139 (1981).

60. N. Daumas and A. Herold, Relations between the step concept and reaction mechanisms in graphite insertion compounds. *C. R. Acad. Sci.* Paris, *Ser. C 268,* 373 (1969).

61. M. B. Dowell and D. S. Badorrek, Diffusion coefficients of Br_2, HNO_3 and $PdCl_2$ in graphite, *Carbon 16,* 241 (1978).

62. E. McRae, D. Billaud, J. F. Mareche, and A. Herold, Basal plane resistivity of alkali metal-graphite compounds, *Physica B+C* (Amsterdam) *99,* 489 (1980).

63. M. E. Misenheimer and H. Zabel, In situ X-ray investigation of stage transformations in potassium-graphite intercalation compounds, *Phys. Rev. B 27,* 1443 (1983).

64. J. M. Thomas, G. R. Millward, R. Schlogl, and H. P. Boehm, Direct imaging of a graphite intercalate. Evidence of inter-penetration of "stages" in graphite ferric chloride, *Mater. Res. Bull. 15,* 671 (1980).

65. G. Hooley, The intercalation of layered structures, *Carbon 18,* 82 (1980).

66. P. G. de Gennes, *The Physics of Liquid Crystals,* International Series of Monographs on Physics, Oxford University Press, Fair-lawn, New Jersey, 1974.

67. S. A. Safran and D. R. Hamaan, Long range elastic interactions and staging in graphite intercalation compounds, *Phys. Rev. Lett. 42,* 1410 (1979).

68. S. A. Safran and D. R. Hamaan, Coherency strains and staging in graphite intercalation compounds, *Physica B+C* (Amsterdam) *99,* 469 (1980).

69. R. Nishitani, Y. Uno, and H. Suematsu, In situ observation of staging in potassium-graphite intercalation compounds, *Phys. Rev. B 27,* 6572 (1983).

Index

Printed and bound by CPI Group (UK) Ltd, Croydon, CR0 4YY

17/10/2024

01775703-0004